APPLIED STATISTICS ALGORITHMS

Mathematics and its Applications
Series Editor: G. M. BELL, Professor of Mathematics, King's College
(KQC), University of London

Statistics and Operational Research
Editor: B. W. CONOLLY, Professor of Operational Research, Queen
Mary College, University of London

*In preparation

APPLIED STATISTICS
ALGORITHMS

Editors:

P. GRIFFITHS, B.Sc., Ph.D.
Oxford University Computing Service

and

I. D. HILL, D.Sc.
Division of Computing and Statistics
Clinical Research Centre of the Medical Research Council
Harrow, Middlesex

Published by
ELLIS HORWOOD LIMITED
Publishers · Chichester

for
THE ROYAL STATISTICAL SOCIETY
London

First published in 1985
and Reprinted in 1986 by
ELLIS HORWOOD LIMITED
Market Cross House, Cooper Street, Chichester, West Sussex, PO19 1EB, England
The publisher's colophon is reproduced from James Gillison's drawing of the ancient Market Cross, Chichester.

Distributors:

Australia and New Zealand:
Jacaranda-Wiley Ltd., Jacaranda Press,
JOHN WILEY & SONS INC.
GPO Box 859, Brisbane, Queensland 4001, Australia

Canada:
JOHN WILEY & SONS CANADA LIMITED
22 Worcester Road, Rexdale, Ontario, Canada

Europe and Africa:
JOHN WILEY & SONS LIMITED
Baffins Lane, Chichester, West Sussex, England

North and South America and the rest of the world:
Halsted Press: a division of
JOHN WILEY & SONS
605 Third Avenue, New York, NY 10158, USA

© 1985 Royal Statistical Society/Ellis Horwood Limited

British Library Cataloguing in Publication Data
Griffiths, Paul
Applied statistics algorithms. —
(Ellis Horwood series in mathematics and its applications)
1. Mathematical statistics   2. Algorithms
I. Title   II. Hill, I.D. (Ian David), *1926–*
001.4'22   QA276
Library of Congress Card No. 85–891

ISBN 0–470–20184–3 (Halsted Press)

Typeset by Ellis Horwood Limited.
Printed in Great Britain by Unwin Brothers of Woking.

Table of Contents

To the memory of
Abu Ja'far Mohammed ben Musa
from whose description as
al-Khwārazmi (the native of Khwārazm)
the word Algorithm is derived

A'LGORISM,   } n. f. Arabick words, which are ufed to imply
A'LGORITHM: } the fix operations of arithmetick, or the fci-
ence of numbers.   *Dict.*

Dr Samuel Johnson's definition of algorithm has been reproduced in facsimile from his
Dictionary of the English Language, (1755)

Foreword

It gives me great satisfaction to introduce this collection of algorithms from *Applied Statistics*. It comes with a scholarly introduction describing the history of algorithms, which stretches back to a time long before a group of us began the algorithm section of *Applied Statistics* in 1968. We began tentatively, and we made inevitable mistakes, some of which are tactfully alluded to in the book; what I suspect none of us realized at the beginning were the difficulties that beset the production of a good algorithm. It is hard to get the code right, to ensure that the algorithm is efficient and well structured, that it adheres to the language standard, that all kinds of potential misuse are trapped, and that the description is clear and accurate. The whole exercise is iterative and, as the editors show, not even the scrutiny of past editors and referees (to whom all of us, users and authors alike, owe so much) has avoided further scrutiny and amendment in preparing the algorithms for reproduction in book form. (Doubtless some energetic individuals will soon be doing their best to prove that all is still not yet perfect — I predict that they will have a hard job.)

Statisticians have become major users of packages, and packages depend heavily on reliable and tested algorithms. The writer of a one-off program for a particular analysis is equally in need of such modules as the algorithms represent; they may be black boxes but the contents must be of top quality. To the extent that increasing numbers of people are now learning statistics from running programs rather than from reading books, the executable algorithm has become part of a new and powerful form of literature, with its own style and perhaps eventually its own masterpieces. The editors have made their careful selection — can you see which may eventually become classics?

J. A. Nelder
President (1985/86)
The Royal Statistical Society
London·

The algorithms contained in this book are available in machine-readable form from the publisher. Please address all enquiries to:

Ellis Horwood Limited, Market Cross House, Cooper Street, Chichester, England, PO19 1EB

Preface

A police electronics engineer admitted in the Johannesburg Magistrates' Court yesterday that, by applying unerring logic, a winning number could be determined in an electronic computerised card game played in a Hillbrow casino. . . . Col R.D. Hull said the electronic computerised machines have an algo rhythm, which is a series of rhythmic manipulations to determine the outcome of a game. The algo rhythm is a "mathematical recipe for numbers" which is the key to the game. The programmer knows the key.

Johannesburg Star
(Acknowledgement to *Computer Weekly* where we first saw this quoted.)

Applied Statistics, one of the journals published by the Royal Statistical Society, has included an Algorithms Section for the last 16 years, and the total number of algorithms published exceeds 200. The time has come for an anthology of them, which we now present.

Algorithms are the building blocks of computer programs, each designed to perform a specific purpose, and to be slotted into a program at will. At its best an algorithm should be so well tested, known to be efficient and always to give the correct results, that it can be used simply as a 'black box' whose inner workings can be ignored, and need not be understood, provided that one thoroughly understands its purpose, how to feed it with input and how to handle its output. But even the blackest of boxes needs its construction to be fully specified so that further copies of it can be made, and so that experts can suggest improvements. It is the aim of both the journal and this book that the algorithms presented should fulfil all such requirements.

As more fully described in Chapter 1, the book does not simply reproduce what has previously appeared in the journal. Corrections and improvements have been made, some previously published as separate entities but now fully incorporated, others new to the book, if further study has shown them to be desirable.

All who write computing instructions know the fascination of pitting one's wits against the pure logic of a machine that does what you told it, not what you meant to tell it. It is humbling as well as fascinating, and we have been humbled often enough to be aware that our aim, for which we have tried so hard, of a book that is free from 'bugs' has almost certainly not been achieved. We ask readers to let us know of any discovered.

ALGORISMS AND ALGORITHMS

Dr R A Griffiths and Mr W R Good (Radcliffe Infirmary, Oxford OX2 6HE) and Mr J G Griffith (Jesus College, University of Oxford) write: We note the use in a recent series of articles of the neologism 'algorithm'. The *Oxford English Dictionary* describes this as a 'recent pseudo-etymological perversion' in which 'algorism' is learnedly confused with 'ἀριθμός, 'number'. 'Algorism' is, of course, from the Arabic and without Greek derivation. 'Algorithm' could only be a fusion of 'ἄλγος (pain) and 'ἀριθμός (number) and would presumably denote 'measurement or counting of pain'.

(*British Medical Journal*, 23 June 1984)

There are many words in the English language that started life as a 'recent pseudo-etymological perversion'. The exact point at which any such word passes the boundary, and becomes correct usage, is difficult to determine, but we have no doubt at all that *algorithm* did so many years ago, and it would be inexcusable affectation to refuse to use it.

Can any word still be a neologism if it appeared in Dr Johnson's dictionary, admittedly not with the current meaning, but with both spellings allowed? He knew only one spelling for 'arithmetick', but Boswell's hope that 'the authority of the great Master of our language will stop that curtailing innovation, by which we see *critic, public,* &c., frequently written instead of *critick, publick,* &c.' remained unsatisfied. Those who think 'algorithm' an incorrect spelling must therefore regard 'arithmetic' as even more so.

Since Dr Johnson's day 'algorithm' has come to mean a recipe for computing operations. Even if it is this meaning that is objected to as neologistic, it goes back nearly half a century at least, as the Supplement to the *Oxford English Dictionary* shows by quoting a reference to 'Euclid's algorithm' from 1938.

Algorism has never taken on this new meaning but remains defined as 'The Arabic, or decimal, system of numeration'. Why should anyone be expected to take an almost unknown word with the wrong meaning, when a better-known word with the right meaning is available?

UNSELECTED ALGORITHMS

To those algorithm authors whose contributions have not been selected for the book we should like to give assurance that lack of space is the reason. No-one should assume that the absence of a given algorithm implies in any way that it lacks merit.

ACKNOWLEDGEMENTS

We thank not only all who have helped in the preparation of this book, but also all whose labours have contributed to the Algorithms Section of the journal. In particular John Nelder as founder and first editor, Howard Simpson as a subsequent editor, and Patrick Royston and Janet Webb who have now taken on the task, deserve gratitude. Maria de la Hunty, as the Society's Executive Editor (for all its journals) has always been a tower of strength. None of them, nor ourselves when in the editorial chair, could have done much without the selfless efforts of all the anonymous referees who so carefully examine and test all the contributions. We are deeply grateful to all those algorithm authors (and their employers in some instances) who have so willingly agreed to let us reproduce their work, and also to the Association for Computing Machinery for permission to use one algorithm from their *Communications*.

Finally, we thank our good friends the computers, but for whom our subject matter would not exist other than as a speculative dream of science fiction.

P. Griffiths
I. D. Hill

December 1984

1
Introduction

1.1 The Royal Statistical Society and its journals

This book puts together, and updates, a number of algorithms previously published in *Applied Statistics,* one of the journals of the Royal Statistical Society.

The Statistical Society of London was founded in 1834, and became the Royal Statistical Society in 1887. Its journal was founded in 1838, at first as a monthly publication but subsequently quarterly. This is now known as *JRSS Series A*. In 1934 a Supplement to the Journal was started. This lapsed in 1941 but was resumed in 1946, and in 1948 became *JRSS Series B*. It specializes in the mathematical theory of statistics.

A third journal, called *Applied Statistics,* commenced publication in 1952. Originally it was published, in conjunction with the Society, by Oliver and Boyd Ltd, but in 1964 the Society took over publication itself, and gave it the additional title of *JRSS Series C*. The name *Applied Statistics* remains with it, however, and is still its title for purposes of giving references, etc. Its aims were set out in a foreword to the first issue as: 'to meet the needs of all workers concerned with statistics — not of professional statisticians only but also of those innumerable workers in industry, commerce, science, and other branches of daily work, who must handle and understand statistics as part of their tasks. Its aim, in short, is to present, in one way or another but always simply and clearly, the statistical approach and its value, and to illustrate in original articles modern statistical methods in their everyday applications'. It therefore concentrates on applications of the subject rather than on theory, which is well served by its companion journals. A volume of *Applied Statistics,* consisting of about 300 — 350 pages, is published each year in three parts.

1.2 Statistical computing

It can be claimed that at least some members of the Society were interested in computing machinery right from the start, for one of its principal founders, at whose house initial committee meetings were held, and who took the chair until the first President was elected, was Charles Babbage, now well known for his designing an Analytical Engine which would have contained most of the features of a computer, with the exception of electrical operation – everything was to be done mechanically with cog-wheels and levers. The first drawing for this proposed machine is dated September 1834, when the Statistical Society was six months old.

His earlier machine, the Difference Engine, had been designed to calculate mathematical tables, but not as a general-purpose computer. A pilot version of this had been built, and became operational in 1822.

Following Babbage's failure to realize his dreams, many people regarded any attempt at computing machinery as doomed to fail. Some remained hopeful however. Thus William Farr (1871) in his Presidential Address to the Society (much of which consisted of an obituary of Babbage) felt 'persuaded that ere many years an analytical machine will be at work'.

A more fruitful line at that time was not automatic machinery, but hand-operated calculating machines. Some of these are mentioned in the *Journal*. For example, W. S. Jevons (1878) says that 'With the Arithmometer at hand, the work becomes rather amusement than labour, especially to those at all fond of ingenious and beautiful mechanism'. At the same meeting the President – G. J. Shaw Lefevre M. P. – says 'Although it might not save time, it would save a great deal of mental labour, and that was one of the greatest considerations in a Government department'.

In 1911 the *Journal* contains an interesting account of twenty or so machines displayed at a reception, and the 1915 volume contains a review of a book on calculating machines, but in general there is little discussion of them at meetings; they are merely a tool in the background.

Meanwhile, outside the auspices of the Society, a computing laboratory formed part of the organization of Karl Pearson's department at University College London. This consisted, not of course of computers in the modern sense, but of people operating hand-cranked calculating machines (mostly of the *Brunsviga* variety) and turning out a splendid collection of tables designed to be useful to statisticians. Some of the tables that they produced formed a series called *Tracts for Computers,* but again the word 'computers' refers to people, not to machines.

Modern problems concerning the copying of software, and the photocopying of publications, lay a long way off, yet even as early as 1914, we find Pearson complaining that, while his tables are published at a loss (with the aid of a charitable grant), he gets accused of 'checking the progress of science' if he tries to stop unauthorized reproductions. 'It is a singular phase of modern science that it steals with a plagiaristic right hand while it stabs with a critical left' (Pearson, 1914).

During the 1920s and 1930s many statistical techniques were invented, such as the fitting of models by maximum likelihood, that were agreed (at least by most statisticians) to be the right way to do things if only the arithmetical labour were not too heavy to be practicable. Nowadays we use these methods regularly with little thought of the amount of computation involved.

The arrival of the Elliott 401 computer at Rothamsted in 1954 can probably be taken as the starting point for modern statistical computing. During these last 30 years the computer has become such a powerful friend of the statistician that it is difficult to imagine what life could be like without it.

1.3 The publication of algorithms

The publication of algorithms, in the sense of methods of performing calculations for particular purposes, is of course much older than electronic computing. The famous textbook of Whittaker and Robinson (1924) is one good example of the way such things used to be done. Even algorithms designed for automatic computing machinery were published in the first half of the nineteenth century (Menabrea, 1842; Lovelace, 1843), showing how Babbage's Analytical Engine could be programmed for specific calculations.

Babbage, incidentally, had been very forward-looking in this, as in so many other ways. He originally planned that if his engine required a logarithmic or trigonometric value it would ring a bell, print a message to its operator asking for the value, and stop. The operator would then give it the value on a pre-punched card. It would check whether it had been given the right value, and if so continue, but otherwise ring a louder bell until satisfied. The checking was possible because the pre-punched card would contain the argument value as well as the result (but he does not seem to have thought of the possibility of getting the right argument for the wrong function!). However, in his autobiography (Babbage, 1864) he adds 'It will be an interesting question, which time only can solve, to know whether such tables of cards will ever be required for the Engine. Tables are used for saving the time of continually computing individual numbers. But the computations to be made by the Engine are so rapid that it seems most probable that it will make shorter work by computing directly from proper formulae than by having recourse even to its own tables'. His hunch was right, and today's computers calculate values from suitable formulae as required rather than store lengthy tables.

So far as we are aware, the first journal to undertake the publication of algorithms for electronic computers, as a regular feature of its contents, was *Communications of the Association for Computing Machinery*, whose Algorithm Number 1 appeared in February 1960. This series was transferred to a new journal, also published by ACM, *Transactions on Mathematical Software*, in 1975. The first 220 *CACM* algorithms were published as received without refereeing, but subsequent ones have all been refereed.

Other journals that entered the field in the first half of the 1960s included *BIT, Numerische Mathematik,* and *Computer Bulletin* whose series was subsequently transferred to *Computer Journal*. The *Applied Statistics* algorithms did not start until 1968.

1.4 The Chilton meeting, and the Working Party on Statistical Computing

A meeting on 'Statistical Programming — the present situation and future prospects' was held at the Atlas Computer Laboratory, Chilton, on 15 December 1966, organized by J. A. Nelder and B. E. Cooper with 39 other participants. The papers read and a brief account of the discussion were subsequently published (Nelder and Cooper, 1967) and now make fascinating reading. As a result of this meeting, a working party was formed to carry matters further. Its initial members were J. A. Nelder (Chairman), B. E. Cooper (Secretary), J. M. Chambers, J. M. Craddock, E. S. Page, J. C. Gower and M. J. R. Healy. The chairman remarked that this was to be a *working* party not a working *party*. It started as an informal group but, at a later date, became officially recognized as a working party of the Society.

The report of its first meeting says that 'The working party is considering how a library of tested algorithms for statistical work could be built up, published and maintained. The example set by the Association for Computing Machinery in publishing algorithms in a standard form is one that deserves close study. If the algorithms in such a library are to be widely useful they would inevitably be required to satisfy rules laid down in the interests of standardization. The specification and development of these rules is important because they will be the foundation on which all other work will depend'.

1.5 Algorithms in *Applied Statistics*

Steps were then taken to form an algorithms section of *Applied Statistics*. A formal policy statement appeared in Volume 17, issue 1 (Anon, 1968) and the first algorithms in issue 2 of that volume, preceded by an article on 'The construction and description of algorithms' by the Working Party, of which some of the membership had changed (Working Party, 1968). Subsequent versions of this article have appeared at intervals, as seemed necessary to keep in step with revised ideas and with computing developments. The latest updated version appears as Chapter 2 of this book.

J. A. Nelder, already a joint editor of the journal, became the first algorithm editor, and got the series off to an excellent start, with a formal layout of introductory text preceding the computer instructions of each and with the latter reproduced photographically from computer listings in the hope of thereby avoiding typesetting errors. On the whole this hope has been fulfilled, and *Applied Statistics* algorithms have become noted for their accuracy. Some of the coding has, or course, contained bugs that authors, editors and referees have all failed to spot, but there have been very few errors introduced by printers, who do their best, but are unfamiliar with the niceties (not to mention the 'nasties') of computer languages. Most users of published algorithms will be familiar with some of the disasters that have appeared in some other journals.

Believing imitation to be the sincerest form of flattery, we have noticed with pleasure that the *Applied Statistics* layout of introductory text has sometimes occurred in print elsewhere. Muxworthy (1982) in discussing program documentation standards says 'the *Applied Statistics* rules are comprehensive, easily available and serve as a good model'.

J. A. Nelder, having edited the algorithms from 1968 until 1972, was succeeded in the post by I. D. Hill (1972–1976), H. R. Simpson (1976–1979), P. Griffiths (1979–1984) and J. P. Royston and J. B. Webb jointly (1984–). Each has had excellent back-up from the Executive Editor at the Society's offices and from a pool of algorithm referees, who help the editors to judge the submissions and to knock them into shape.

Deciding that there were too many algorithms in different journals called Algorithm 1, Algorithm 2, etc., it was decided in the first place to include the initials of the journal and use the designations Algorithm AS 1, Algorithm AS 2, etc. An unexpressed hope (until now) that other journals would take the hint and preface theirs with suitable initials has not been fulfilled.

1.6 Languages and standards

The algorithms section of the journal has sometimes been criticized for being too backward-looking in having published only in Fortran (1966 vintage), with an occasional leavening of Algol 60, and one solitary example of PL/I. This is largely a question of authors wishing their algorithms to be compatible with others, so that a number may be used within one program; but the more authors see that Fortran is the main thing published the more they get the impression that Fortran is the only thing acceptable. So only Fortran gets submitted, and the editors cannot publish what does not get submitted.

At least one of us believes that the world took a disastrously wrong step, back in the 1960s and 1970s, in going so heavily for Fortran (and Cobol in the commercial field) when the Algol path was available. Some of this language battle is already visible in the report of the Chilton meeting already mentioned — and it was one of the original members of the working party who (in private conversation) once said 'If it hasn't been published in Fortran, it hasn't been published'. The present widespread use, in the micro-computer field, of the language called Basic (which pretends to be easy, but is in fact unnecessarily difficult for non-trivial programs) can be regretted by both Fortran and Algol addicts. Basic is not acceptable to *Applied Statistics* until there is a standard version that includes self-contained routines.

It has to be admitted that, however regrettable, in one sense Fortran won the battle, becoming universally available while Algol is now uncommon. For that reason the Algol algorithms in this book are accompanied by Fortran translations. There was a temptation to include Algol translations of the Fortran ones as well, to show how much better some of them could be done, but it had to be admitted that the extra pages (and hence extra cost) would not have been worth it. For one thing, those who most need to be shown better things would not have looked at them.

In another sense, however, Algol won the battle in that many developments to Fortran (and Basic and other languages), some already made and others proposed, whether as non-standard extensions or as revised and amended standard versions, are merely implementing features that have been in Algol since 1960. Slowly and painfully they are introducing such features, but rarely as well done as in the original. Again we detect imitation as the sincerest form of flattery, though not *all* features of Algol 60 are worth imitating.

Also, in recent years, Pascal has acquired a considerable following although no Pascal algorithm has yet been accepted by *Applied Statistics*. The Algol lover cannot help regretting some features of Pascal, but it has succeeded in getting a language of the Algol family widely recognized, and the best of its features are excellent.

From the start, *Applied Statistics* has been keen that its algorithms should be in accordance with the appropriate language standards. However, in the early days, there was some confusion about what the Fortran standard said, and many users assumed that the standard agreed with the features of the particular compiler that they normally used. Thus, of the seven algorithms published in the first bunch, six are in Fortran and each is said to be in 'ASA Standard Fortran' (ASA was the old name of ANSI — the American National Standards Institute), yet only one of the six is actually in accordance with that standard (and that one was spoilt by the printer, who deleted the *C* in column 1 of comment lines when preparing the photographed listing!)

Similar trouble is likely, if great care is not taken, if and when authors start submitting in Fortran 77. Recently one of our colleagues was expressing amazement that, although *END IF* is part of the standard, *END DO,* which he regularly uses on a particular machine, is not. People may be submitting these non-standard extensions before long.

Over the years the references to this language were changed from ASA Fortran to ANSI Fortran, and then to ISO Fortran when the International Standardization Organization produced their own version of it which, it was decided, should be considered as the definitive one. Now that Fortran 77 has become both the ANSI and the ISO standards, a further change has become necessary, to Fortran 66, which has never been an official name but will be universally understood. It is surprising the extent to which the 1966 version is still taken as being *the* standard. Even 6 years after it ceased to be so we find references such as by Henstridge *et al.* (1984) which says 'The programs as presented are designed for compilation under standard ANSI Fortran [with a reference to the 1966 standard]. They may be compiled under a Fortran 77 [reference] compiler . . . etc.'.

1.7 Remarks and corrigenda
The third bunch of algorithms also included the first in a series of remarks (designated AS R1, AS R2, etc.) and corrigenda on algorithms already published. Such items are more common with algorithms than with research papers, not because algorithm authors are more careless but because: (a) errors matter more — a spelling mistake in plain English text may be unsightly, but will rarely cause significant harm, but a similar mistake in computer code is disastrous; (b) errors are more likely to be discovered when algorithms are run by different people using different data, different machines and different compilers.

Such remarks and corrigenda are, inevitably, not published together with the algorithms to which they refer, but in a later issue. Consequently users often fail to discover them and copy an unamended original version. One reason for this book is therefore to incorporate corrections into the chosen algorithms to give corrected and updated versions.

1.8 Planning this book

A number of vague suggestions have been made at one time and another for a book such as this, usually by people who thought that it could be done by throwing together existing pages, with no real work involved. The first firm suggestion was made by the *Applied Statistics* editorial board at their meeting in December 1979, asking for a proposal to be drawn up and presented to their next meeting in April 1980.

A request was then circulated to 29 interested people, asking them to vote on which algorithms should be included. Each algorithm, from AS 1 to AS 156, was to be classified as: (i) minimal set; (ii) also essential; (iii) desirable; (iv) undesirable. In the event what emerged from the 17 replies was far from perfect agreement, but clear preferences for those areas of statistics in which algorithms were required did emerge, and the voting helped a great deal in forming the set actually presented here. In addition to the voting figures we paid particular attention to completeness and consistency, in the sense that if algorithm A is included and it makes use of algorithm B, then that also is included. This policy extends to one algorithm from another collection (ACM 291).

It has been the aim, so far as possible, to take algorithms from the first hundred or so, on the grounds that it will now be known whether they are useful or not, and that any bugs in them will probably have been found and documented. However, some later ones have been taken where they are, in effect, improved versions of older ones that had initially been selected, and one (AS 183) because there was a particular request (not from its authors) for its inclusion. There were' also some surprises among the older ones in new bugs being discovered unexpectedly. On the whole we believe the aim to have been met that the algorithms in this book, having been updated and corrected from their originally published versions, are reliable and accurate; but obviously no guarantee can be given that they are bug-free.

The updating has included the incorporation of all known and published corrections and improvements, together with some unpublished ones that further study has shown to be desirable. Additional changes have been made to improve uniformity and conformity to the latest *Applied Statistics* standards. The introductory text to each algorithm contains a section headed 'Editors' Remarks' detailing what has been done.

References
Anon. (1968) Statistical algorithms — editorial note. *Appl. Statist.*, **17**, 79–82.
Babbage, C. (1864) *Passages from the Life of a Philosopher*. London: Longman, Green, Longman, Roberts and Green.
Farr, W. (1871) President's inaugural address. *J. Statist. Soc.*, **34**, 409–423.
Henstridge, J. D., Payne, R. W. and Baker, R. J. (1984) Algorithm 117 — Buffered output in Fortran. *Comput. J.*, **27**, 179–182.

Jevons, W. S. (1878) On the statistical use of the Arithmometer. *J. Statist. Soc.*, **41**, 597–599.

Lovelace, A. A. (1843) Article XXIX. Translation of 'Sketch of the Analytical Engine' with notes by the translator. *Taylor's Scientific Memoirs*, iii, 666–731.

Menabrea, L. F. (1842) Sketch of the Analytical Engine invented by Charles Babbage Esq. *Bibliotheque Universelle de Genève*, No. 82.

Muxworthy, D. T. (1982) Program documentation standards. *SSRC Data Archive Bulletin* No. 21.

Nelder, J. A. and Cooper, B. E. (1967) Statistical programming (including papers by other authors, and discussion). *Appl. Statist.*, **16**, 87–151.

Pearson, K. (1914) Preface to *Tables for Statisticians and Biometricians*, Part 1. Cambridge: Cambridge University Press.

Whittaker, E. T. and Robinson, G. (1924) *The Calculus of Observations*. London and Glasgow: Blackie & Son Ltd.

Working Party on Statistical Computing (1968) The construction and description of algorithms. *Appl. Statist.*, **17**, 175–179.

2

The construction and description of algorithms

Revised by P. Griffiths, I. D. Hill, J. P. Royston and J. B. Webb

2.0 Introduction

Notes on the form to be taken by contributions to the Algorithms Section of *Applied Statistics* have appeared previously under this title (Working Party on Statistical Computing, 1968, 1975, 1979). This revision follows closely the recommendations issued in 1979, but reflects the experience gained and the changes in computing practice which have taken place in the intervening years.

2.1 Submission

Algorithms for *Applied Statistics* should be submitted to the Secretary, Royal Statistical Society, and consist of three copies of each of: (1) the introductory text; (2) the algorithm itself; (3) a driver program, not for publication but for the use of the referee, which should call the algorithm and produce checkable output. It should test both usual and unusual cases. Items (2) and (3) should be in the form of computer listings.

Authors are not expected to supply their algorithms in computer-readable form as a matter of course, but referees are entitled to ask for such, and authors should be ready to supply it if requested.

Submissions in the form of short notes or remarks on previously published algorithms are also invited.

2.2 Refereeing

Adequate refereeing of an algorithm can entail much computer testing, so that refereeing may take longer than for a paper of comparable length. However, speed must be regarded as secondary to thoroughness.

The author is welcome to submit additional information (e.g. a flow diagram) to assist the referee.

The Algorithm Editors welcome volunteers to join the pool of referees.

2.3 Languages

Algorithms are published for two purposes: for direct use and for communicating computing methods. In either case the value of the algorithm is dependent upon its being written in a well-defined language for ease of both comprehension and implementation.

So far nearly all algorithms have been in Fortran, to the 1966 Standard (which we shall call Fortran 66) (ANSI, 1966; ISO, 1972), with some in Algol 60 (ISO, 1984). Most of the Fortran 66 algorithms are also valid in the current Fortran Standard (Fortran 77 – ANSI, 1978). Algorithms specifically in Fortran 77, using features not to be found in Fortran 66, are also acceptable. Other languages that are regarded as sufficiently standardized for the purpose, and have an adequate procedure structure, include Algol 68 (van Wijngaarden *et al.*, 1976), Pascal (BSI, 1982), etc. Authors using these or other languages should be aware of the difficulties which may be encountered in finding referees familiar with both the statistical content and the language used, and with access to a relevant compiler. Basic is not, at present, acceptable because it lacks a standard version that includes the concept of independent procedures.

Authors should also consider whether they wish their algorithms to be used in conjunction with others that have been published, in which case there may be a clear case for using Fortran even if it would not otherwise be the language of choice.

It is probably wise not to use the more esoteric features of some languages (such as parallel processing) which are not widely implemented, and the inclusion of which would severely handicap refereeing and subsequent use of the algorithm. If any libraries or separately compiled modules are used, appropriate comment must be made in the Description, and a reference provided to enable potential users to check the details; this may also cause problems for referees.

In general we interpret the language standards very rigidly, and refuse non-standard features even when widely implemented. It is important that the algorithms should run successfully using any compiler that satisfies the appropriate standard. Only a rigid interpretation can ensure this. (See also sections 2.19 and 2.20 below.)

The driver program, which is not for publication, need not follow the language standard quite so rigidly provided that it is easily understandable and implementable by the referee. It must be remembered, however, that the referee may be using a different model of computer (indeed it is preferable that he should) and may never even have heard of the non-standard feature that the author uses regularly. Wise authors do not upset referees by putting them to needless trouble.

2.4 Introductory text

The general form of this can be seen by looking at the algorithms in a recent issue of the journal. They normally contain most of the following sections, as applicable.

TITLE.

AUTHOR. Author's name and affiliation.

KEYWORDS. A list of headings suitable for an information retrieval service. For full details see Anon. (1971).

LANGUAGE. See section 2.3 above.

DESCRIPTION AND PURPOSE. This should enable the potential user to decide whether the algorithm is likely to serve the purpose, and referees and others to judge it technically. Reference to published work should be made when this will shorten the description. Subheadings are optional; the following may be useful: *Purpose; Notation; Theory; Numerical method.*

STRUCTURE. This gives the heading of the algorithm, so that the user can see at a glance whether it is a subroutine or a function, its identifier and its list of formal parameters. This heading is followed by subsections dealing with formal parameters, failure indications, auxiliary algorithms, etc. as appropriate.

Formal parameters. These should be listed in the same order as in the heading, giving for each its type, bounds if an array, its classification as input and/or output, workspace or value (if 'call by value' is a relevant feature of the language used) together with a brief description of its purpose. A parameter may be listed as both input and output when appropriate, with a separate brief description for each purpose. If a parameter is listed as input only, the algorithm must not under any circumstances change its value.

Failure indications. Types of failure should be listed and described (see also section 2.11 below).

Auxiliary algorithms. This should give references to any published algorithms required as auxiliaries. When suitable, algorithms in the *Applied Statistics* series should be quoted as first choice, while those in widely available journals are preferable to those published in obscure places. References to a subroutine library that is available only to users of computers from a particular manufacturer will not be entertained, since they limit the transportability of the resulting algorithm. Preferable by far is the provision of a precise description of the requirements, supplemented by an indication of any suitable routines in widely available subroutine libraries. If an auxiliary algorithm is published with a main algorithm, but may sometimes be called directly by a user, then its structure should be specified as for a main algorithm. If however, the auxiliary is intended to be called only by the main algorithm, then its structure need not be specified in detail.

Constants. This should list any constants that a user might wish to change, to attain different accuracy or to be suitable for a computer of different word-length, explaining how the values were chosen and how to modify them.

RESTRICTIONS. Restrictions on the use of an algorithm should be given. Examples are an upper limit to the size of a matrix, or the inability of an algorithm to deal with missing observations. So far as possible restrictions on the size of the problem that can be handled should be avoided by using dynamic storage within the routine, even at the cost of additional parameters in Fortran (see also sections 2.17 and 2.18 below).

PRECISION. This is a most difficult point, since it is desired that the algorithms should be as machine-independent as possible. Yet an algorithm may require 'double precision' when run on a machine that uses only 32 bits to represent real variables, whereas it would be absurd to use 'double precision' for the same algorithm on a machine that uses 48 or 60 bits. The rule for *Applied Statistics* algorithms is that, unless a special case can be made (for example, the need for two different degrees of precision within one algorithm), only single precision can be accepted, but that a section of the introductory text should describe the changes needed for a double-precision version (see also section 2.9 below).

TIME. Where possible, run times should be given in units relatively independent of the machine configuration. Failing this, times for running problems of various sizes with a given machine configuration may be quoted.

ACCURACY. The accuracy of the results should be related to the accuracy of the data, the word length of the computer and the method.

RELATED ALGORITHMS. Where relevant, the possibilities of combining the algorithm with other published algorithms should be commented on, and if other published algorithms exist for the same purpose, comparisons must be made.

TEST DATA. A short set of test data, with expected results, can be very useful in determining whether an algorithm has been correctly implemented. Such data should be designed to be as revealing as possible.

ADDITIONAL COMMENTS. Comments on the best way to use the algorithm may be included here. For example, it may be possible to save running time by programming an inner loop in machine code, or by including as an integral part of the algorithm an operation usually done by an auxiliary.

REFERENCES. Publications should be cited in the conventional form for *Applied Statistics*.

2.5 Layout

To reduce errors, algorithms are reproduced from computer files, after being passed through editing programs which produce a standard layout. Authors, therefore, need not worry about the exact details of layout, but that submitted must of course be in accordance with the requirements of the language, and should be neat enough to make the referee's job easy.

In a language where hardware representations differ from the Reference Language, the algorithm, being a computer listing, must necessarily be in hardware representation. Any unusual representation should be explained to help the referee. The translation to Reference Language for publication is part of the function of the editing program.

In Algol the editing program insets the text to make easy the pairing of each **begin** with its corresponding **end**, but initial submission in indented form can often make the task of the referee more palatable. No such insetting is used for Fortran 66, although some programmers like to indicate the *DO*-loop structure in this way. There are two reasons for not adopting such insetting: (1) that an inset labelled Fortran statement cannot have its label inset, so that several levels of insetting lead to statements becoming more and more divorced from their labels; (2) the printed page, with none of the guidelines of a coding form, must have the position of column 6 clearly defined. This is best achieved by starting all statements in column 7, with insetting used only for continuation lines where column 6 is clearly indicated.

In Fortran 77, the use of a *BLOCK-IF* structure makes insetting more desirable, and also leads to fewer labels being used. The above considerations therefore apply less strongly, and insetting may well be adopted when appropriate.

2.6 Comments

Every algorithm must have an opening comment, which should give a brief description of its purpose. However, this should not include lengthy descriptions of formal parameters, references, etc., since these details are given in the printed text. Further comments should be used liberally to make human understanding easier. In Algol 60 lengthy comments should always be introduced by the **comment** symbol. Comments following **end** should be restricted to a brief indication of what is being ended.

2.7 Input/output

Unless input or output is of the essence of the algorithm, these operations should be avoided. The operation of the algorithm, in general, should be purely internal leaving the user to decide on any interaction with the external world.

Algol 60 input/output must be by means of procedure calls, not by using pseudo-Algol 60 basic symbols (such as **read**, **print** or **format**) such as are supported by some compilers. Such procedures can be regarded as auxiliary to the main algorithm, but a concise description of their functions must be included within the Description (see section 2.4 above).

2.8 Constants and variables

Constants, whose value may be machine-dependent, are distinguished from formal parameters in that they can be given new values only by recompiling the algorithm. Constants denoting 'a very small number' or 'a very large number' are examples. Such local constants should be referred to symbolically and their values isolated in the first few statements with explanatory comment; the Fortran *DATA* statement is useful here. Constants such as $\sqrt{(2\pi)}$ or the coefficients of a polynomial may also with advantage be set in the same way. In Fortran 77, the *PARAMETER* statement may be preferable to the *DATA* statement for such a purpose.

Real constants should be written with at least one digit both before and after the point. Thus 1.0 and 0.5 are preferred to 1. and .5

When using mixed-mode arithmetic (not allowed in Fortran 66) constants used in a real context should be expressed in real form.

2.9 Precision

In Fortran (and other languages that allow different precisions), authors are strongly advised: (1) to declare all *REAL* identifiers in a type-statement; (2) to denote all *REAL* constants symbolically, setting values by a *DATA* or *PARA-METER* statement; (3) to enter standard functions through statement functions. Thus

$$REAL\ A, B, C, HALF, ONE, PI, ZSQRT$$
$$DATA\ HALF, ONE, PI/0.5, 1.0, 3.141593/$$
$$ZSQRT\ (A) = SQRT\ (A)$$

enables the user to change just these three lines to read

$$DOUBLE\ PRECISION\ A, B, C, HALF, ONE, PI, ZSQRT$$
$$DATA\ HALF, ONE, PI/0.5D0, 1.0D0, 3.14159\ 265359D0/$$
$$ZSQRT\ (A) = DSQRT\ (A)$$

without needing to hunt through the algorithm for places where changes are needed.

2.10 Machine-dependence

Machine-dependent algorithms, such as one assuming a particular sequence of characters in the internal character code, must be avoided.

Methods of access to backing store, such as discs and magnetic tapes, are also machine-dependent, and the arrangement of information in these stores may depend on the source of the information, which could be another program. If storage and retrieval operations are segregated, preferably in subsidiary algorithms, the parent algorithm will be more easily adapted to particular uses. The functions of such subsidiary algorithms must be defined in the specification and their position marked by program comment.

In Fortran 77 some of these things have become more standard, and may therefore be used more freely in algorithms than before.

2.11 Error diagnostics

Algorithms must detect errors and be protected against misuse. Where the main purpose of an algorithm is output, the inclusion of an error message among that output may be appropriate. In other cases it is recommended that output of diagnostic information should be avoided. Instead the formal parameter list of every algorithm in which a fault is possible should include an integer error indicator *IFAULT* (in Fortran) or *ifault* (in Algol). This should normally be set to zero by the routine, but if a fault occurs the algorithm should reset *IFAULT* to values 1,2,3, . . . indicating the type of error. The treatment of faults which can be detected initially can be handled in a variety of ways; in some circumstances the following approach can be useful:

$$IFAULT = 1$$
$$IF(. . . .)\ RETURN$$
$$IFAULT = 2$$
$$IF(. . . .)\ RETURN$$
$$IFAULT = 0$$

If feasible, default values may be set following a fault, so that the algorithm can continue.

If the algorithm is in the form of a function, it must give a value to the function, in addition to setting *IFAULT,* before returning. It is common to give the value zero in such circumstances.

Users of algorithms should be expected to test the value of *IFAULT* after calling the algorithm (unless faults are impossible in the particular context of use).

When an algorithm calls an auxiliary algorithm that has an *IFAULT* parameter, it must test for a fault and take appropriate action if necessary, or have a comment to explain why the fault cannot occur. The final setting of the main algorithm's own fault parameter (if any) should be sufficient to indicate what all the errors were and in which algorithms they occurred.

Where possible the fundamental, rather than immediate, causes of errors should be listed. For example, a negative square root might indicate a non-positive definite matrix, or division by zero (or by a very small number) might arise when inverting a singular matrix. Any storage limitations should be checked; users might not always be in a position to make these checks themselves, and if the algorithm is unprotected the reasons for a program failure can often be difficult to determine.

2.12 Parentheses
Unnecessary brackets in expressions should usually be avoided. Though their inclusion may have little effect in terms of machine efficiency, they can make it more difficult for the human reader to establish the meaning of the expression.

2.13 Labels, jumps and ifs
The flow of control through an algorithm should be kept as simple as possible, without unnecessary labels and jumping. Fortran cannot manage without labels, but these should be in numerical order (not necessarily consecutive numbers) to make it easier to follow the operations.

Algorithms in structured languages can often be written without any labels at all, and generally should be, but where a **goto** is the simplest, neatest solution it should be adopted in preference to a highly involved structure which avoids using **goto**. In particular, an *EXIT* label at the end allows **goto** *EXIT* to mimic the Fortran *RETURN* statement, and may well avoid an unnecessarily elaborate **begin end** structure.

In Fortran 66, the simplicity of

$$Y = A$$
$$IF (I.EQ. 0) Y = B$$

is usually preferable to

$$IF (I.EQ. 0) GOTO 50$$
$$Y = A$$
$$GOTO 60$$
$$50 \quad Y = B$$
$$60 \ldots$$

even though it means an unnecessary assignment to Y if $I = 0$. In Fortran 77

```
IF (I .EQ. 0) THEN
      Y = B
ELSE
      Y = A
ENDIF
```

is the preferred construction, and some other languages can handle this sort of thing more neatly.

There are many other examples of clumsy constructions that can be handled more neatly if a little thought is applied.

We prefer logical-*IF* to arithmetic-*IF* because its meaning is so much clearer. An arithmetic-*IF* is acceptable if the choice is genuinely between going to three different labels, but a construction such as

```
      IF (J) 5, 5, 10
  5 ...
```

is much better written as

```
      IF (J .GT. 0) GOTO 10
```

2.14 Construction of loops
Neither Fortran nor Algol 60 is particularly good at indicating to the human reader when the end of a loop is reached, neither having any equivalent of **od** in Algol 68, or *NEXT* in Basic.

Some indication is advisable whenever the body of a loop is a compound statement in Algol 60 or more than one statement in Fortran. In Algol 60 this can be a comment following the **end**. In Fortran the terminal statement of a *DO* should be *CONTINUE* unless the body of the loop is only one statement (and this statement is not a logical-*IF*). The *CONTINUE* statement should not be used elsewhere.

Wherever possible, operations should be performed once outside a loop rather than many times inside. It should never be assumed that an optimizing compiler will be available to handle this.

2.15 Parameters in Algol 60
A good general rule is that simple variables should be called by value unless they must be by name, but arrays by name unless they must be by value.

Some Algol 60 programmers do not realize that the great generality of its call by name (which can be very useful in the right circumstances) means that the referencing of a simple variable by name may require considerable computation to find its value, or its address, according to context. To use such a parameter as a working variable, where an internally declared variable would do equally well, is absurdly inefficient.

2.16 Common storage and global identifiers

Procedures should be self-contained, avoiding global identifiers and Common storage. However, when a set of procedures is designed to be used as a package, global identifiers or labelled Common may be used as a means of communication between procedures within the package; use of unlabelled Common will not be accepted.

2.17 Use of work-space

When the work-space requirements of an algorithm can be calculated from parameters of the problem, the amount should be given algebraically in the specification.

When work arrays in Algol procedures are declared internally, the compiler will allocate storage. When a Fortran algorithm is considered in isolation, the usual practice is to allocate such arrays for the private use of the algorithm. The storage requirements of a set of algorithms used in the same program is the sum of the requirements of the components, unless a work-space array with name *WS* (say) is declared in the main program, and included as an argument for all algorithms that require it. The work-space required for the main program will then be the maximum required for any individual algorithm.

When a Fortran algorithm requires more than one work-space array, different starting points *IA, IB, IC, . . .* (say) within a single array *WS* may be assigned to the different arrays. These starting points can be evaluated either in the main program, when the call to the subroutine will include arguments *WS(IA)*, *WS(IB)*, *WS(IC)*, . . . or within the subroutine, when only a single argument *WS* occurs in the parameter list, and reference, for example, to *WS(K)*, where $K = IB + J$, would refer to the $(J+1)$th store in the second work array.

2.18 Multidimensional arrays

Some statistical algorithms operate on multidimensional arrays with variable numbers of dimensions. This situation is most conveniently tackled by declaring one-dimensional arrays and programming the multidimensional aspects. This need not result in an inefficient program, for advantage can be taken of any systematic order in which the array elements are required. Skill in this type of simulation is soon acquired and is also useful when the number of dimensions allowed by the compiler is limited.

In particular, if one of the dimensions of a two-dimensional array is never used as a variable, but reference to the array is always in the form *A(1,J)* or *A(2,J)*, then the use of two separate one-dimensional arrays, say *A1(J)* and *A2(J)*, is preferable.

When an algorithm is published to communicate a technique, efficiency is less important than clarity and multidimensional forms are recommended. In other cases the alternative multidimensional form may be given as comment.

2.19 Standard Fortran

Many Fortran compiler-writers seem to regard it as a virtue to include extensions to the language. For an amusing example of just how far extensions can go, see Haddon (1975). Few algorithm authors would dare to submit such an example, but inadvertent departures from the standard are frequent. The most common in Fortran 66 are as follows: (1) The use of *DIMENSION A*(1) where *A* is a formal parameter, and the given array bound is purely a dummy value. The use of the element *A*(2), or above, is then illegal. Either a numerical value must be given, in which case it must not exceed the equivalent value in the calling routine, or a variable bound must be given such as *DIMENSION A*(*N*) where *N* is also in the parameter list. In neither case may a subscript value exceed the given bound; (2) The use of apostrophes instead of the *H* notation for character strings; (3) The use of characters, in comment, that are not within the Fortran character set; (4) The use of negative or zero values, or of expressions, in the parameters of a *DO*; (5) Mixed-mode arithmetic, in which real and integer values both occur in an expression (other than *real ** integer* which is allowed); (6) Subscript expressions that do not fall within the very limited range of forms allowed; even *A*(*I+J*) is not permissible; (7) The use of an unsubscripted array name in a *DATA* statement. Each array element must be listed individually; (8) The assumption that a local variable, other than one initialized in a *DATA* statement and not subsequently altered in value, keeps its value from subroutine exit to the next entry.

Fortran 77 is less restrictive and allows most of these constructions. For the first one, *DIMENSION A*(*) may be used for the purpose, and for the last the *SAVE* statement is available.

However, neither version allows *REAL*8 instead of *DOUBLE PRECISION*, while both insist that if two dummy arguments both take the same actual argument, then assignment to either of them is disallowed.

2.20 Character handling

The Fortran 66 standard allows any type of variable to hold a 'Hollerith' (character) value, but it does not allow such a value to be assigned from one variable to another, nor does it allow a comparison of two variables holding such values.

We allow 'Hollerith' values of not more than four characters to be held in real variables, and not more than two characters in integers. They may be assigned to other variables of the same type only, and be compared with others of the same type for equality or inequality only. If such operations are performed, however, they should be mentioned in the introductory text, as not all compilers will handle them satisfactorily.

Fortran 77 uses *CHARACTER* type instead, and does not suffer from these restrictions.

2.21 Copyright and disclaimer

The copyright of all contributions published in the Algorithm Section remains with the Royal Statistical Society. Requests for permission to reproduce any material should be made to the Secretary, Royal Statistical Society.

While every effort is made to ensure the accuracy and efficiency of algorithms published, no liability is assumed by any contributor or by the Royal Statistical Society, its officers or *Applied Statistics*.

References

Anon. (1971) Notes for the preparation of mathematical papers. *J. R. Statist. Soc. B*, **33**, following p. 330.

ANSI (1966) American National Standard FORTRAN, X 3.9 − 1966.

ANSI (1978) American National Standard FORTRAN, X 3.9 − 1978.

BSI (1982) Specification for computer programming language Pascal. BS6192: 1982.

Haddon, E. W. (1975) Book review of *Digital Computing and Numerical Methods. Computer J.*, **18**, 69.

ISO (1972) Recommendation R 1539, *Programming Language* FORTRAN.

ISO (1984) International Standard 1538, *Programming Language* ALGOL 60.

Van Wijngaarden, A., Mailloux, B. J., Peck, J. E. L., Koster, C. H. A., Sintzoff, M., Lindsey, C. H., Meertens, L. G. L. T. and Fisker, R. G. (eds) (1976) *Revised Report on the Algorithmic Language Algol 68*. Berlin: Springer-Verlag.

Working Party on Statistical Computing (1968) The construction and description of algorithms. *Appl. Statist.*, **17**, 175−179.

Working Party on Statistical Computing (1975) The construction and description of algorithms. *Appl. Statist.*, **24**, 366−373.

Working Party on Statistical Computing (1979) The construction and description of algorithms. *Appl. Statist.*, **28**, 311−318.

The Algorithms

Algorithm AS 3

THE INTEGRAL OF STUDENT'S *t*-DISTRIBUTION
By B. E. Cooper
Atlas Computer Laboratory

Present address: London School of Economics, Houghton Street, London WC2A 2AE, UK.

Keywords: Student's distribution; *t*-distribution; Tail area.

LANGUAGE

Fortran 66 and 77.

DESCRIPTION AND PURPOSE

This function computes the area from $-\infty$ to t_ν under a Student's central *t*-distribution with ν degrees of freedom.

Numerical method
The numerical method used is obtained as a special case of a method given by Owen (1965) for a non-central *t*-distribution as used and described in Algorithm AS 5.

STRUCTURE

REAL FUNCTION PROBST (T, IDF, IFAULT)

Formal parameters

T	Real	input:	the value of t_ν.
IDF	Integer	input:	the degrees of freedom ν.
IFAULT	Integer	output:	1 if *IDF* < 1; 0 otherwise.

PRECISION

For a double precision version, change *REAL FUNCTION* to *DOUBLE PRECISION FUNCTION, REAL* to *DOUBLE PRECISION, SQRT* to *DSQRT, ATAN* to *DATAN,* and give double precision versions of the constants in the *DATA* statement.

ACCURACY

The method is theoretically exact. Inaccuracies due to rounding or cancellation are such that the result should be accurate to 11 decimal places on a computer working to 12 significant decimal figures.

RELATED ALGORITHMS

The same results can be obtained using Algorithm AS 5 by setting the non-centrality parameter to zero. However, the present subroutine is shorter and quicker and possibly more accurate.

EDITORS' REMARKS

The algorithm remains exactly as originally published in principle. The style has been modified to meet current *Applied Statistics* standards. A modification has been made to prevent 'underflow' on large degrees of freedom.

REFERENCE

Owen, D. B. (1965) A special case of a bivariate non-central *t*-distribution. *Biometrika*, **52**, 437–446.

```
      REAL FUNCTION PROBST(T, IDF, IFAULT)
C
C         ALGORITHM AS 3   APPL. STATIST. (1968) VOL.17, P.189
C
C         STUDENT T PROBABILITY (LOWER TAIL)
C
      REAL A, B, C, F, G1, S, FK, T, ZERO, ONE, TWO, HALF, ZSQRT, ZATAN
C
C         G1 IS RECIPROCAL OF PI
C
      DATA ZERO, ONE, TWO, HALF, G1
     $     /0.0, 1.0, 2.0,  0.5, 0.3183098862/
C
      ZSQRT(A) = SQRT(A)
      ZATAN(A) = ATAN(A)
C
      IFAULT = 1
      PROBST = ZERO
      IF (IDF .LT. 1) RETURN
      IFAULT = 0
      F = IDF
      A = T / ZSQRT(F)
      B = F / (F + T ** 2)
      IM2 = IDF - 2
      IOE = MOD(IDF, 2)
      S = ONE
      C = ONE
      F = ONE
      KS = 2 + IOE
      FK = KS
      IF (IM2 .LT. 2) GOTO 20
      DO 10 K = KS, IM2, 2
      C = C * B * (FK - ONE) / FK
      S = S + C
      IF (S .EQ. F) GOTO 20
      F = S
      FK = FK + TWO
   10 CONTINUE
   20 IF (IOE .EQ. 1) GOTO 30
      PROBST = HALF + HALF * A * ZSQRT(B) * S
      RETURN
   30 IF (IDF .EQ. 1) S = ZERO
      PROBST = HALF + (A * B * S + ZATAN(A)) * G1
      RETURN
      END
```

Algorithm AS 5

THE INTEGRAL OF THE NON-CENTRAL t-DISTRIBUTION

By B. E. Cooper

Atlas Computer Laboratory

Present address: London School of Economics, Houghton Street, London WC2A 2AE, UK.

Keywords: Non-central t-distribution; Tail area.

LANGUAGE

Fortran 66 and 77.

DESCRIPTION AND PURPOSE

This function computes the area from $-\infty$ to t_0 under a non-central t-distribution with ν degrees of freedom and non-centrality parameter δ.

Numerical method

The numerical method closely follows that given by Owen (1965). The integration is theoretically exact given that the auxiliary functions provide exact values. For degrees of freedom exceeding 100, a normal approximation is used.

STRUCTURE

REAL FUNCTION PRNCST (ST, IDF, D, IFAULT)

Formal parameters

ST	Real	input:	the value of t_0.
IDF	Integer	input:	the degrees of freedom ν.
D	Real	input:	the non-centrality parameter δ.
IFAULT	Integer	output:	0 if exact method used;
			1 if normal approximation used;
			2 if $IDF < 1$.

Auxiliary algorithms

The following auxiliary routines are called:

REAL FUNCTION ALNORM (X, UPPER) – Algorithm AS 66;
REAL FUNCTION TFN (X, FX) – Algorithm AS 76;
REAL FUNCTION ALOGAM (X, IFAULT) – Algorithm ACM 291.

PRECISION

For a double precision version, change *REAL FUNCTION* to *DOUBLE PRECISION FUNCTION*, *REAL* to *DOUBLE PRECISION*, *SQRT* to *DSQRT*,

EXP to *DEXP*, use double precision versions of the auxiliary algorithms, and give double precision versions of the constants in the *DATA* statements.

ACCURACY

The accuracy of the probability obtained is determined by the accuracy of *TFN* (Algorithm AS 76). The result should be accurate to more than six decimal places, except when *IFAULT* = 1.

EDITORS' REMARKS

The method is the same as originally published, but the auxiliary routines have been replaced with later algorithms for the same purposes. Remark AS R8 (Hitchin, 1973) is incorporated. The style has been updated. A normal approximation has been introduced for large degrees of freedom, for which the original algorithm fails. A modification has been incorporated to guard against overflow under some circumstances, following a remark recently submitted by Youn-Min Chou.

REFERENCES

Hitchin, D. (1973) Remark AS R8. *Appl. Statist.*, **22**, 428.
Owen, D. B. (1965) A special case of a bivariate non-central *t*-distribution. *Biometrika*, **52**, 437–446.

```
      REAL FUNCTION PRNCST(ST, IDF, D, IFAULT)
C
C         ALGORITHM AS 5  APPL. STATIST. (1968) VOL.17, P.193
C
C         COMPUTES LOWER TAIL AREA OF NON-CENTRAL T-DISTRIBUTION
C
      REAL ST, D, G1, G2, G3, ZERO, ONE, TWO, HALF, EPS, EMIN, F,
     $ A, B, RB, DA, DRB, FMKM1, FMKM2, SUM, AK, FK, FKM1,
     $ ALNORM, TFN, ALOGAM, ZSQRT, ZEXP
C
C         CONSTANTS - G1 IS 1.0 / SQRT(2.0 * PI)
C                     G2 IS 1.0 / (2.0 * PI)
C                     G3 IS SQRT(2.0 * PI)
C
      DATA G1, G2, G3 /0.3989422804, 0.1591549431, 2.5066282746/
      DATA ZERO, ONE, TWO, HALF,    EPS, EMIN
     $     /0.0, 1.0, 2.0,  0.5, 1.0E-6, 12.5/
C
      ZSQRT(A) = SQRT(A)
      ZEXP(A) = EXP(A)
C
      F = IDF
      IF (IDF .GT. 100) GOTO 50
      IFAULT = 0
      IOE = MOD(IDF, 2)
      A = ST / ZSQRT(F)
      B = F / (F + ST ** 2)
      RB = ZSQRT(B)
```

```
        DA = D * A
        DRB = D * RB
        SUM = ZERO
        IF (IDF .EQ. 1) GOTO 30
        FMKM2 = ZERO
        IF (ABS(DRB) .LT. EMIN) FMKM2 = A * RB * ZEXP(-HALF * DRB ** 2)
     $    * ALNORM(A * DRB, .FALSE.) * G1
        FMKM1 = B * DA * FMKM2
        IF (ABS(D) .LT. EMIN)
     $  FMKM1 = FMKM1 + B * A * G2 * ZEXP(-HALF * D ** 2)
        IF (IOE .EQ. 0) SUM = FMKM2
        IF (IOE .EQ. 1) SUM = FMKM1
        IF (IDF .LT. 4) GOTO 20
        IFM2 = IDF - 2
        AK = ONE
        FK = TWO
        DO 10 K = 2, IFM2, 2
        FKM1 = FK - ONE
        FMKM2 = B * (DA * AK * FMKM1 + FMKM2) * FKM1 / FK
        AK = ONE / (AK * FKM1)
        FMKM1 = B * (DA * AK * FMKM2 + FMKM1) * FK / (FK + ONE)
        IF (IOE .EQ. 0) SUM = SUM + FMKM2
        IF (IOE .EQ. 1) SUM = SUM + FMKM1
        AK = ONE / (AK * FK)
        FK = FK + TWO
   10 CONTINUE
   20 IF (IOE .EQ. 0) GOTO 40
   30 PRNCST = ALNORM(DRB, .TRUE.) + TWO * (SUM + TFN(DRB, A))
        RETURN
   40 PRNCST = ALNORM(D, .TRUE.) + SUM * G3
        RETURN
C
C         NORMAL APPROXIMATION - K IS NOT TESTED AFTER THE TWO CALLS
C         OF ALOGAM, BECAUSE A FAULT IS IMPOSSIBLE WHEN F EXCEEDS 100
C
   50 IFAULT = 1
        A = ZSQRT(HALF * F) * ZEXP(ALOGAM(HALF * (F - ONE), K))
     $    - ALOGAM(HALF * F, K)) * D
        PRNCST = ALNORM((ST - A) / ZSQRT(F * (ONE + D ** 2)
     $    / (F - TWO) - A ** 2), .FALSE.)
        RETURN
        END
```

Algorithm AS 6

TRIANGULAR DECOMPOSITION OF A SYMMETRIC MATRIX
By M. J. R. Healy
MRC Clinical Research Centre

Present address: London School of Hygiene and Tropical Medicine, Keppel Street, London WC1E 7HT, UK.

Keywords: Triangular decomposition; Symmetric matrix; Cholesky.

LANGUAGE

Fortran 66 and 77.

DESCRIPTION

Any positive semi-definite symmetric matrix $\mathbf{A}$ can be expressed in the form $\mathbf{A} = \mathbf{U}^T\mathbf{U}$, where $\mathbf{U}$ is a real upper triangular matrix. If $\mathbf{A}$ is singular some of the diagonal elements of $\mathbf{U}$ (in number equal to the nullity of $\mathbf{A}$) are zero; the other elements of $\mathbf{U}$ in the corresponding rows are then taken to be zero. Given $\mathbf{A}$, the subroutine produces the matrix $\mathbf{U}$.

The algorithm is based on the following formulae

$$u_{ii}^2 = a_{ii} - \sum_{j=1}^{i-1} u_{ji}^2 \qquad u_{ii}u_{ik} = a_{ik} - \sum_{j=1}^{i-1} u_{ji}u_{jk}, \; k > i.$$

In practice, u_{ii} (and all u_{ik}) are set equal to $0 \cdot 0$ if

$$|w_i| < ETA \times |a_{ii}|$$

where

$$w_i = a_{ii} - \sum_{j=1}^{i-1} u_{ji}^2$$

and the local constant ETA is a pre-set tolerance.

STRUCTURE

SUBROUTINE CHOL(A, N, NN, U, NULLTY, IFAULT)

Formal parameters

A	Real array (NN)	input:	the input matrix, stored as a one-dimensional array in the sequence $a_{11}, a_{21}, a_{22}, a_{31}, a_{32}, a_{33}, \ldots$
N	Integer	input:	the order of A.

NN	Integer	input:	the size of the A and U arrays $(N(N + 1)/2)$.
U	Real array (*NN*)	output:	the result matrix, stored as a one-dimensional array in the sequence $u_{11}, u_{12}, u_{22}, u_{13}, u_{23} \ldots$
NULLTY	Integer	output:	the nullity of A; hence the number of u_{ii} that have been set to zero.
IFAULT	Integer	output:	a fault indicator equal to: 1 if N is less than 1; 2 if A is not positive semi-definite; 3 if $NN \neq N(N + 1)/2$; 0 otherwise.

Constant

The local constant *ETA* should be set according to the precision of the arithmetic. 10^{-5} is about right for 32-bit representation, 10^{-9} for 48-bit or 64-bit.

RESTRICTIONS

None. Some compilers will allow U to coincide with A and the algorithm is so written that no harm will be done, but this is illegal in Standard Fortran.

PRECISION

For a double precision version, change *REAL* to *DOUBLE PRECISION, ABS* to *DABS, SQRT* to *DSQRT,* and give double precision versions of the constants in the *DATA* statement. The value of *ETA* should be changed as mentioned above.

EDITORS' REMARKS

This algorithm has been modified to incorporate the suggestions of Farebrother and Berry (1974) and Barrett and Healy (1978). The style has been modified to meet current *Applied Statistics* standards.

The further modifications of Freeman (1982), to allow submatrices to be inverted more efficiently, have not been included, but readers' attention is drawn to them.

REFERENCES

Barrett, J. C. and Healy, M. J. R. (1978) Remark AS R27. *Appl. Statist.,* **27,** 379–380.

Farebrother, R. W. and Berry, G. (1974) Remark AS R12. *Appl. Statist.,* **23,** 477.

Freeman, P. R. (1982) Remark AS R44. *Appl. Statist.,* **31,** 336–339.

```
      SUBROUTINE CHOL(A, N, NN, U, NULLTY, IFAULT)
C
C        ALGORITHM AS 6  APPL. STATIST. (1968) VOL.17, P.195
C
C        GIVEN A SYMMETRIC MATRIX ORDER N AS LOWER TRIANGLE IN A( )
C        CALCULATES AN UPPER TRIANGLE, U( ), SUCH THAT UPRIME * U = A.
C        A MUST BE POSITIVE SEMI-DEFINITE. ETA IS SET TO MULTIPLYING
C        FACTOR DETERMINING EFFECTIVE ZERO FOR PIVOT.
C
      REAL A(NN), U(NN), ETA, ETA2, X, W, ZERO, ZABS, ZSQRT
C
C        THE VALUE OF ETA WILL DEPEND ON THE WORD-LENGTH OF THE
C        COMPUTER BEING USED. SEE INTRODUCTORY TEXT.
C
      DATA ETA, ZERO /1.0E-5, 0.0/
C
      ZABS(X) = ABS(X)
      ZSQRT(X) = SQRT(X)
C
      IFAULT = 1
      IF (N .LE. 0) RETURN
      IFAULT = 3
      IF (NN .NE. N * (N + 1) / 2) RETURN
      IFAULT = 2
      NULLTY = 0
      J = 1
      K = 0
      ETA2 = ETA * ETA
      II = 0
      DO 80 ICOL = 1, N
      II = II + ICOL
      X = ETA2 * A(II)
      L = 0
      KK = 0
      DO 40 IROW = 1, ICOL
      KK = KK + IROW
      K = K + 1
      W = A(K)
      M = J
      DO 10 I = 1, IROW
      L = L + 1
      IF (I .EQ. IROW) GOTO 20
      W = W - U(L) * U(M)
      M = M + 1
   10 CONTINUE
   20 IF (IROW .EQ. ICOL) GOTO 50
      IF (U(L) .EQ. ZERO) GOTO 30
      U(K) = W / U(L)
      GOTO 40
   30 IF (W * W .GT. ZABS(X * A(KK))) RETURN
      U(K) = ZERO
   40 CONTINUE
   50 IF (ZABS(W) .LE. ZABS(ETA * A(K))) GOTO 60
      IF (W .LT. ZERO) RETURN
      U(K) = ZSQRT(W)
      GOTO 70
   60 U(K) = ZERO
      NULLTY = NULLTY + 1
   70 J = J + ICOL
   80 CONTINUE
      IFAULT = 0
      RETURN
      END
```

Algorithm AS 7

INVERSION OF A POSITIVE SEMI-DEFINITE SYMMETRIC MATRIX

By M. J. R. Healy

MRC Clinical Research Centre

Present address: London School of Hygiene and Tropical Medicine, Keppel Street, London WC1E 7HT, UK.

Keywords: Matrix inversion; Cholesky; Generalized inverse; G-inverse.

LANGUAGE

Fortran 66 and 77.

DESCRIPTION

Let $\mathbf{A}$ be an $n \times n$ positive semi-definite symmetric matrix of rank $r \leqslant n$. A generalized inverse of $\mathbf{A}$ (Rao, 1962) can be defined as follows. Apply the same permutation to the rows and columns of $\mathbf{A}$ in such a way as to produce a symmetric matrix

$$\begin{pmatrix} \mathbf{A}_{11} & \mathbf{A}_{12} \\ \mathbf{A}_{12}^{\mathrm{T}} & \mathbf{A}_{22} \end{pmatrix}$$

where $\mathbf{A}_{11}$ is a non-singular $r \times r$ principal minor. Apply the reverse permutation to the rows and columns of the matrix

$$\begin{pmatrix} \mathbf{A}_{11}^{-1} & \mathbf{0} \\ \mathbf{0} & \mathbf{0} \end{pmatrix}$$

where the $\mathbf{0}$s denote matrices of zeros. The result ($\mathbf{A}^-$ say) is a generalized inverse of $\mathbf{A}$ under Rao's definition. In particular, if the linear equations $\mathbf{A}\mathbf{x} = \mathbf{h}$ are consistent, then $\mathbf{x} = \mathbf{A}^-\mathbf{h}$ is a set of solutions.

The numerical method involves expressing $\mathbf{A}$ in the form $\mathbf{U}^{\mathrm{T}}\mathbf{U}$ with $\mathbf{U}$ upper triangular and forming a generalized inverse of $\mathbf{U}$. The detection of singularity is governed by a tolerance parameter *ETA* set inside *CHOL*.

STRUCTURE

SUBROUTINE SYMINV (A, N, NN, C, W, NULLTY, IFAULT)

Formal parameters

A	Real array (*NN*)	input:	the input matrix, stored as a one-dimensional array in lower triangular sequence $a_{11}, a_{21}, a_{22}, a_{31}, a_{32}, a_{33}. \ldots$

N	Integer	input:	the order of A.
NN	Integer	input:	the size of the A and C arrays $(N(N + 1)/2)$.
C	Real array (NN)	output:	the output matrix, stored like A.
W	Real array (N)	workspace:	
$NULLTY$	Integer	input:	the nullity of A (and of C).
$IFAULT$	Integer	output:	a fault indicator, equal to:
			1 if N is less than 1;
			2 if A is not positive semi-definite;
			3 if $NN \neq N(N + 1)/2$;
			0 otherwise.

Auxiliary algorithm
The subroutine *CHOL* (Algorithm AS 6) is used.

RESTRICTIONS

None. Some compilers will allow C to coincide with A and the algorithm is so written that no harm will be done, but this is illegal in Standard Fortran.

ACCURACY

Approximations to segments of the Hilbert matrix were inverted and the results re-inverted. The computer used was the ICT Atlas (48-bit word). For $N = 5$, the largest relative discrepancy was $0 \cdot 0000063$; for $N = 10$, the largest relative discrepancy was $0 \cdot 26$; for $N = 15$, the matrix was declared not to be positive semi-definite. The matrix with

$$a_{ii} = N + 1; \quad a_{ij} = N, \quad i \neq j$$

was inverted and the result re-inverted. Up to $N = 15$, the original matrix was reproduced to 9 significant figures.

I am indebted to Messrs B. E. Cooper and A. J. H. Walter for these tests.

PRECISION

For a double precision version, change *REAL* to *DOUBLE PRECISION*, and give double precision versions of the constants in the *DATA* statement.

EDITORS' REMARKS

This algorithm remains as originally published in principle. The style has been modified to meet current *Applied Statistics* standards. The modifications of Freeman (1982), to allow submatrices to be inverted more efficiently, have not been included, but readers' attention is drawn to them.

REFERENCES

Rao, C. R. (1962) A note on a generalized inverse of a matrix with application
    to problems in mathematical statistics. *J. R. Statist. Soc. B,* **24,** 152–158.
Freeman, P. R. (1982) Remark AS R44. *Appl. Statist.,* **31,** 336–339.

```
      SUBROUTINE SYMINV(A, N, NN, C, W, NULLTY, IFAULT)
C
C         ALGORITHM AS 7   APPL. STATIST. (1968) VOL.17, P.198
C
C         FORMS IN C( ) AS LOWER TRIANGLE, A GENERALIZED INVERSE
C         OF THE POSITIVE SEMI-DEFINITE SYMMETRIC MATRIX A( )
C         ORDER N, STORED AS LOWER TRIANGLE.
C
      REAL A(NN), C(NN), W(N), X, ZERO, ONE
C
      DATA ZERO, ONE /0.0, 1.0/
C
      CALL CHOL(A, N, NN, C, NULLTY, IFAULT)
      IF (IFAULT .NE. 0) RETURN
      IROW = N
      NDIAG = NN
   10 L = NDIAG
      IF (C(NDIAG) .EQ. ZERO) GOTO 60
      DO 20 I = IROW, N
      W(I) = C(L)
      L = L + I
   20 CONTINUE
      ICOL = N
      JCOL = NN
      MDIAG = NN
   30 L = JCOL
      X = ZERO
      IF (ICOL .EQ. IROW) X = ONE / W(IROW)
      K = N
   40 IF (K .EQ. IROW) GOTO 50
      X = X - W(K) * C(L)
      K = K - 1
      L = L - 1
      IF (L .GT. MDIAG) L = L - K + 1
      GOTO 40
   50 C(L) = X / W(IROW)
      IF (ICOL .EQ. IROW) GOTO 80
      MDIAG = MDIAG - ICOL
      ICOL = ICOL - 1
      JCOL = JCOL - 1
      GOTO 30
   60 DO 70 J = IROW, N
      C(L) = ZERO
      L = L + J
   70 CONTINUE
   80 NDIAG = NDIAG - IROW
      IROW = IROW - 1
      IF (IROW .NE. 0) GOTO 10
      RETURN
      END
```

Algorithm AS 13

MINIMUM SPANNING TREE
By G. J. S. Ross
*Rothamsted Experimental Station,
Harpenden, Herts, AL5 2JQ, UK.*

Keywords: Minimum spanning tree; Distance matrix; Similarity matrix.

LANGUAGES
Algol 60, and Fortran 66 and 77.

DESCRIPTION AND PURPOSE

This procedure computes the minimum spanning tree of a distance matrix by the method of Prim, as described by Gower and Ross (1969). The procedure can be easily modified to handle a similarity matrix by reversing the relevant inequalities. The distance matrix is assumed stored in lower triangular form with diagonal terms omitted. The procedure can be modified to handle matrices stored in other ways.

D is a lower triangular distance matrix with diagonal elements omitted. *dlarge* is a constant, larger than any element of D. The object of the procedure is to define for each number $2 \leqslant i \leqslant n$ a partner $1 \leqslant B_i \leqslant n$ with corresponding distance C_i such that the tree is connected and of minimum length. The results are thus the arrays B and C. In some languages B and C, and the internal array A, may be conveniently packed as a single array. Unknown distances can be handled if they are coded as *dlarge*. The procedure can equally well be adapted for use with similarity matrices (minimum distance = maximum similarity, *dlarge* = 0, etc.).

STRUCTURE

procedure *primtree (n, dlarge, D, B, C, ifault)*

Formal parameters

n	Integer	value:	the number of points = the order of the matrix.
dlarge	Real	value:	an arbitrary value larger than any element of D. A large power of 2 is usually suitable.
D	Real array $[1:n \times (n-1)/2]$	input:	the lower triangular distance matrix.
B	Integer array $[2:n]$	output:	$B[i]$ contains the index of a point to which i is jointed.

C	Real array $[2{:}n]$	output:	$C[i]$ is the distance between i and $B[i]$.
ifault	Integer	output:	set to: 1 if $n < 2$; 0 otherwise.

RESTRICTIONS

The minimum permissible value of n is 2. The maximum value will be determined by storage considerations.

TIME

The time required depends on n^2.

ACCURACY

The resulting tree may not be unique if the matrix contains equal elements.

EDITORS' REMARKS

Minor changes have been made in the text. The algorithm itself remains as originally published except that the A array is made **Boolean** instead of **integer**. A Fortran version is added for compatibility with other algorithms in this book.

REFERENCES

Gower, J. C. and Ross, G. J. S. (1969) Minimum spanning trees and single linkage cluster analysis. *Appl. Statist.*, **18**, 54–64.
Prim, R. C. (1957) *Bell System Tech. J.*, **36**, 1389–1401.

procedure *primtree*(*n, dlarge, D, B, C, ifault*);

comment Algorithm AS 13, Appl. Statist., (1969), Vol.18, p.103;

value *n, dlarge*; **integer** *n, ifault*; **real** *dlarge*;
real array *D, C*; **integer array** *B*;

comment Computes the minimum spanning tree of a distance matrix;

    begin
    integer *i, j, k, next*; **real** *min, dist*;
    Boolean array *A*[2 : *n*];

    comment *A*[*i*] is **false** if *i* is already assigned to the tree
    (initially consisting of no. 1 only), or **true** otherwise;

```
if n > 1 then
    begin
    ifault := 0;
    for i := 2 step 1 until n do
        begin
        A[i] := true; B[i] := 0;
        C[i] := dlarge
        end i loop;
    j := 1;
    for i := 2 step 1 until n do
        begin
        min := dlarge;
        for k := 2 step 1 until n do
        if A[k] then
            begin
            dist := D[if j > k then (j − 1) × (j − 2) ÷ 2 + k
                                else (k − 1) × (k − 2) ÷ 2 + j];
            if dist < C[k] then
                begin
                C[k] := dist; B[k] := j
                end;
            if min > C[k] then
                begin
                min := C[k]; next := k
                end
            end k loop;
        j := next; A[j] := false
        end i loop
    end
else ifault := 1
end primtree
```

Fortran version of Algorithm AS 13

STRUCTURE

SUBROUTINE PRTREE (N, M, A, DLARGE, D, B, C, IFAULT)

Formal parameters
N, DLARGE, D, B, and C are similar to the corresponding parameters of the Algol version. The B and C arrays run from 1 to N instead of from 2 to N.

M	Integer	input:	the value of $N(N-1)/2$.
A	Logical array (N)	workspace:	
IFAULT	Integer	output:	1 if $N < 2$;
			2 if $M \neq N(N-1)/2$;
			0 otherwise.

PRECISION

For a double precision version, change *REAL* to *DOUBLE PRECISION*.

```
      SUBROUTINE PRTREE(N, M, A, DLARGE, D, B, C, IFAULT)
C
C        ALGORITHM AS 13   APPL. STATIST. (1969) VOL.18, P.103
C
C        COMPUTES THE MINIMUM SPANNING TREE OF A DISTANCE MATRIX
C
      REAL D(M), C(N), AM, DIST, DLARGE
      INTEGER B(N)
      LOGICAL A(N)
      IFAULT = 1
      IF (N .LT. 2) RETURN
      IFAULT = 2
      IF (N * (N - 1) / 2 .NE. M) RETURN
      IFAULT = 0
C
C        A(I) IS .FALSE. IF I IS ALREADY ASSIGNED TO
C        THE TREE (INITIALLY CONSISTING OF NO. 1 ONLY),
C        OR .TRUE. OTHERWISE
C
      DO 10 I = 2, N
      A(I) = .TRUE.
      B(I) = 0
      C(I) = DLARGE
   10 CONTINUE
      J = 1
      DO 40 I = 2, N
      AM = DLARGE
      DO 30 K = 2, N
      IF (.NOT. A(K)) GOTO 30
      IF (J .GT. K) L = (J - 1) * (J - 2) / 2 + K
      IF (J .LE. K) L = (K - 1) * (K - 2) / 2 + J
      DIST = D(L)
      IF (DIST .GE. C(K)) GOTO 20
      C(K) = DIST
      B(K) = J
   20 IF (AM .LE. C(K)) GOTO 30
      AM = C(K)
      NEXT = K
   30 CONTINUE
      J = NEXT
      A(J) = .FALSE.
   40 CONTINUE
      RETURN
      END
```

Algorithm AS 14

PRINTING THE MINIMUM SPANNING TREE

By G. J. S. Ross

Rothamsted Experimental Station,
Harpenden, Herts, AL5 2JQ, UK.

Keywords: Printing; Output; Minimum spanning tree.

LANGUAGES

Algol 60, and Fortran 66 and 77.

DESCRIPTION AND PURPOSE

This procedure prints the links of the Minimum Spanning Tree in an order which is helpful in preparing it for display. The method is described by Gower and Ross (1969). The procedure assumes that the output of Algorithm AS 13 (*primtree*) is available.

STRUCTURE

procedure *mintreeprint (n, B, C, xprint)*

Formal parameters

n	Integer	value:	the number of points.
B	Integer array [2:*n*]	value:	as defined in **AS** 13.
C	Real array [2:*n*]	input:	as defined in **AS** 13.
xprint	Procedure		see below.

Auxiliary procedure
procedure *xprint (j, k, m, dkm)* prints a new line, *j* spaces, the integers *k* and *m*, and the real number *dkm,* according to local output conventions.

TIME

Time depends on n^2.

EDITORS' REMARKS

Minor changes have been made in the text, and the algorithm has been slightly simplified in form. A Fortran version is added for compatibility with the other algorithms in this book.

The Fortran version presented here is merely a transliteration of the published Algol. Improved ways of achieving the same ends have since been devised and are employed in a Fortran version currently used at Rothamsted Experimental Station. The author can supply details if requested.

REFERENCES

As for Algorithm AS 13.

procedure *mintreeprint*(*n*, *B*, *C*, *xprint*);

comment Algorithm AS 14, Appl. Statist., (1969), Vol.18, p.105;

value *n*, *B*; **integer** *n*; **integer array** *B*;
real array *C*; **procedure** *xprint*;

comment This procedure enables the minimum spanning tree to be drawn rapidly without having to search for end points, and is especially useful when $n > 100$. The output of *primtree* is used;

```
begin integer array hist, route[1 : n]; integer i, j, k, m;
for i := 1 step 1 until n do hist[i] := 0;
for i := 2 step 1 until n do hist[B[i]] := hist[B[i]] + 1;
route[1] := j := k := 1;
```

comment *route*[*j*] is the current end point and if *hist*[*route*[*i*]] $\neq$ 0 a further line can be found. *hist*[1] must be non-zero initially because the tree is connected;

```
for i := 2 step 1 until n do
    begin
    for j := j − 1 while hist[k] = 0 do k := route[j];
    hist[k] := hist[k] − 1;
    for m := 2 step 1 until n do
    if k = B[m] then goto L;
L: xprint(j + 1, k, m, C[m]); j := j + 2;
    route[j] := k := m; B[m] := −B[m]
    end i loop
end mintreeprint
```

Fortran version of Algorithm AS 14

STRUCTURE

SUBROUTINE MTP (N, B, C, HIST, ROUTE, XPRINT)

Formal parameters
N, *B*, *C* and *XPRINT* are similar to the corresponding parameters of the Algol version. The *B* and *C* arrays run from 1 to *N* instead of from 2 to *N*.

HIST Integer array (*N*) workspace:
ROUTE Integer array (*N*) workspace:

Auxiliary procedure
SUBROUTINE XPRINT (J, K, M, DKM) is required to print, on a new line, *J*
spaces, the integers *K* and *M* and the real number *DKM*. A suitable Fortran
format may be devised for this requirement. The actual argument corresponding
to *XPRINT* must be declared as *EXTERNAL* in the routine where *MTP* is called.

PRECISION

For a double precision version, change *REAL* to *DOUBLE PRECISION*. The
fourth argument of *XPRINT* must also be changed similarly.

```
      SUBROUTINE MTP(N, B, C, HIST, ROUTE, XPRINT)
C
C         ALGORITHM AS 14   APPL. STATIST. (1969) VOL.18, P.105
C
C         THIS SUBROUTINE ENABLES THE MINIMUM SPANNING TREE TO BE
C         DRAWN RAPIDLY WITHOUT HAVING TO SEARCH FOR END POINTS,
C         AND IS ESPECIALLY USEFUL WHEN N EXCEEDS 100. THE OUTPUT
C         OF SUBROUTINE PRTREE IS USED
C
      INTEGER B(N), HIST(N), ROUTE(N)
      REAL C(N)
      DO 10 I = 1, N
   10 HIST(I) = 0
      DO 20 I = 2, N
      J = B(I)
      HIST(J) = HIST(J) + 1
   20 CONTINUE
      ROUTE(1) = 1
      J = 1
      K = 1
C
C         ROUTE(J) IS THE CURRENT END POINT AND IF HIST(ROUTE(I))
C         IS NOT ZERO A FURTHER LINE CAN BE FOUND. HIST(I) MUST
C         BE NON-ZERO INITIALLY BECAUSE THE TREE IS CONNECTED
C
      DO 80 I = 2, N
   30 IF (HIST(K) .NE. 0) GOTO 40
      J = J - 1
      K = ROUTE(J)
      GOTO 30
   40 HIST(K) = HIST(K) - 1
      DO 50 M = 2, N
      IF (K .EQ. B(M)) GOTO 60
   50 CONTINUE
   60 CALL XPRINT(J, K, M, C(M))
      J = J + 1
      ROUTE(J) = M
      K = M
      B(M) = -B(M)
   80 CONTINUE
      DO 90 I = 2, N
   90 B(I) = IABS(B(I))
      RETURN
      END
```

Algorithm AS 15

SINGLE LINKAGE CLUSTER ANALYSIS

By G. J. S. Ross

Rothamsted Experimental Station,
Harpenden, Herts, AL5 2JQ, UK.

Keywords: Single linkage; Cluster analysis; Dendrogram.

LANGUAGES

Algol 60, and Fortran 66 and 77.

DESCRIPTION AND PURPOSE

This procedure uses the minimum spanning tree to compute a single linkage cluster analysis as described by Gower and Ross (1969). Two forms of output are provided, (i) a list of the members of each group at each level of clustering and (ii) a dendrogram which summarizes the information in (i).

Single linkage clustering defines, for any given distances threshold *level*, clusters such that any two points of a cluster can be connected by a chain of links each of length less than *level*, a property not possessed by any two points not in the same cluster. Each iteration raises the threshold by *delta* and amalgamates any clusters joined by a single link, thus forming a hierarchical structure which can be represented by a dendrogram. The process terminates when all points lie within a single cluster.

STRUCTURE

procedure *singlelinkage (n, delta, B, C, dlarge, groupprint, topprint, sideprint, printx)*

Formal parameters

n	Integer	value:	the number of points.
delta	Real	value:	the amount by which the clustering threshold is raised at each iteration.
B	Integer array [2:*n*]	input:	as defined in AS 13.
C	Real array [2:*n*]	value:	as defined in AS 13.
dlarge	Real	value:	as defined in AS 13.
groupprint	Procedure		see below.
topprint	Procedure		see below.
sideprint	Procedure		see below.
printx	Procedure		see below.

Auxiliary procedures

(i) **procedure** *groupprint (m, n, G, H)*.

This procedure prints the list, $G[1:n]$, of groups (whose last members are non-zero in $H[1:n]$) formed at level m, according to local output conventions. Something like

print (m);
for $p:= 1$ **step** 1 **until** n **do**
      begin
      print $(G[p])$;
      if $H[p] = 1$ **then** *write* ('*')
      end

is suitable, together with carriage control characters as required.

(ii) **procedure** *topprint* prints topheadings for the dendrogram.

(iii) **procedure** *sideprint(i)* prints a new line and a side title for each line of the dendrogram.

(iv) **procedure** *printx(j)* prints three characters for each element of the dendrogram. These should be as follows:

    $j = 0$ three spaces,
        1 two underlines, space,
        2 two underlines, vertical bar,
        3 three underlines,
        4 two spaces, vertical bar,
        5 two underlines, vertical bar.

TIME

Time depends partly on $n \times$ the number of iterations required to form a single cluster and partly on n^2.

RESTRICTIONS

None, except that if *delta* is too small an unnecessary amount of printing is required. About 10–15 iterations give a satisfactory dendrogram. If more than 20 iterations are required the full dendrogram is not printed, excess portions on the right-hand side being omitted.

COMMENTS

In Algorithms AS 13, AS 14 and AS 15 much space and time can be saved by use of packing techniques.

I am indebted to Mr A. J. H. Walter of Atlas Computer Laboratory, Chilton, who tested these algorithms.

EDITORS' REMARKS

Minor changes have been made to both the text and the algorithm, without in any way changing its effect. A Fortran version is added for compatibility with other algorithms in this book.

The Fortran version presented here is merely a transliteration of the published Algol. Improved ways of achieving the same ends have since been devised and are employed in a Fortran version currently used at Rothamsted Experimental Station. The author can supply details if requested.

REFERENCES

As for Algorithm AS 13.

procedure *singlelinkage(n, delta, B, C, dlarge, groupprint,*
    topprint, sideprint, printx);

comment Algorithm AS 15, Appl. Statist., (1969), Vol.18, p.106;

value *n, delta, C, dlarge;* **integer** *n;* **real** *delta, dlarge;*
integer array *B;* **real array** *C;*
procedure *groupprint, topprint, sideprint, printx;*

comment Performs single linkage clustering. Information is supplied by procedure *primtree*, in arrays *B* and *C*. Points are listed in sorted order in array *G*, and the corresponding array *H* marks the last member of each cluster with a 1, otherwise the entry is 0. The array *X* stores code numbers for output of the dendrogram;

> **begin integer** *i, j, k, m, p, q, r, s, t, u, v, w;* **real** *dmin, level;*
> **integer array** *G, H*[1 : *n*], *X*[1 : 20 × *n*], *W*1, *W*2[0 : *n*];
>
> **comment** Clustering starts at the first integral multiple of *delta* which is greater than *dmin*, the shortest link of the minimum spanning tree;
>
> *dmin* := *C*[2];
> **for** *i* := 3 **step** 1 **until** *n* **do**
> **if** *dmin* > *C*[*i*] **then** *dmin* := *C*[*i*];
> **for** *i* := 1 **step** 1 **until** *n* **do**
>   **begin**
>   *G*[*i*] := *i; H*[*i*] := 1;
>   *X*[*i*] := 3
>   **end** *i loop;*
> *p* := 0;

```
for level := delta × (1 + entier(dmin / delta)),
    level + delta while k ≠ 1 do
        begin

        comment For each link in array C that is shorter than level,
        two clusters are amalgamated. The amalgamation involves a
        reordering of arrays G and H and removal of the end-of-cluster
        marker from the earlier cluster. Links once used are increased
        by dlarge to prevent re-use;

        p := p + 1;
        for i := 2 step 1 until n do
        if C[i] < level then
            begin
            j := B[i]; C[i] := C[i] + dlarge;
            k := i;
            for m := 1 step 1 until n do
                begin
                if G[m] = j then q := m;
                if G[m] = k then r := m
                end m loop;
            if q > r then
                begin
                m := r; r := q;
                q := m
                end;
            for s := q step 1 until n do
            if H[s] ≠ 0 then goto NEXT1;
NEXT1:      for t := r − 1 step −1 until 1 do
            if H[t] ≠ 0 then goto NEXT2;
NEXT2:      t := t + 1; H[s] := 0;
            for r := t step 1 until n do
                begin
                W1[r − t] := G[r]; W2[r − t] := H[r];
                if H[r] ≠ 0 then goto NEXT3
                end r loop;
NEXT3:      w := s + 1; u := r − t + 1;
            for m := t − 1 step −1 until w do
                begin
                G[m + u] := G[m]; H[m + u] := H[m]
                end m loop;
            u := r − t;
            for m := 0 step 1 until u do
                begin
                G[m + w] := W1[m]; H[m + w] := W2[m]
                end m loop
            end i loop;
        groupprint(level, n, G, H);
        w := n × p; u := v := k := 0;
```

comment Dendrogram indicators are now compiled and stored in X. Points are examined in the order defined by the array G. s is the corresponding entry on the previous iteration, $u = 0$ for the first member of a cluster, $v = 1$ when amalgamations occur, from the last member of the first component cluster until the last member of the amalgamated cluster. k is the total number of clusters;

```
for i := 2 step 1 until n do k := k + H[i];
if p < 20 then
for i := 1 step 1 until n do
    begin
    j := G[i]; s := X[j + w − n];
    if u = 0 then
        begin
        if H[i] = 1 then t := 3 else
        if s = 3 then t := u := v := 1 else
            begin
            t := 0; u := 1
            end
        end
    else
    if H[i] = 1 then
        begin
        if v = 0 then
            begin
            t := 3; u := 0
            end
        else
            begin
            t := 2; u := v := 0
            end
        end
    else
    if s = 2 ∨ s = 3 then
        begin
        if v = 0 then t := u := v := 1 else
            begin
            t := 5; u := 1
            end
        end
    else
    if v = 0 then
        begin
        t := 0; u := 1
        end
    else t := 4;
    X[j + w] := t
    end i loop
end level loop;
```

topprint;
for $i := 1$ **step** 1 **until** n **do**
    begin
      $j := G[i]$; *sideprint(j)*;
      if $p > 19$ **then** $p := 19$;
      for $m := 0$ **step** 1 **until** p **do** *printx*($X[m \times n + j]$)
    end i *loop*;

comment Print newlines to separate the dendrogram. Packing in the X store saves considerable space and allows non-significant blank spaces before newline to be recognised and ignored. Packing G and H together simplifies the reordering process;

end *singlelinkage*

Fortran Version of Algorithm AS 15

STRUCTURE

SUBROUTINE SLINK (N, N1, N20, DELTA, B, C, DLARGE, G, H, X, W1, W2, GROUPP, TOPP, PRINTX, IFAULT)

Formal parameters
N, *DELTA*, B, C and *DLARGE* are similar to the corresponding parameters of the Algol version, except that the real array C is not 'called by value' and will be destroyed. If the values are required they must be copied to another array first. The B and C arrays run from 1 to N instead of from 2 to N.

N1	Integer	input:	$N + 1$.
N20	Integer	input:	$20 N$.
G	Integer array (N)	workspace:	
H	Integer array (N)	workspace:	
X	Integer array $(N20)$	workspace:	
W1	Integer array $(N1)$	workspace:	
W2	Integer array $(N1)$	workspace:	
GROUPP	Procedure		see below.
TOPP	Procedure		see below.
PRINTX	Procedure		see below.
IFAULT	Integer	output:	1 if $N1 \neq N + 1$ or $N20 \neq 20 N$; 0 otherwise.

Auxiliary procedures
SUBROUTINE GROUPP (AM, N, G, H) and *SUBROUTINE TOPP* are required, defined similarly to procedures *groupprint* and *topprint* for the Algol version.

SUBROUTINE PRINTX (I, IG, N, IP, IX, N20) is required with integer arguments
I, N, IP and *N20*, integer arrays *IG(N)* and *IX(N20)*. It should perform the
functions of both *sideprint* and *printx* of the Algol version.

PRECISION

For a double precision version, change *REAL* to *DOUBLE PRECISION*, and
give a double precision version of the constant in the *DATA* statement. Also
change *AINT* to *DINT* if the latter is available though non-standard in Fortran
66. Otherwise change *AINT(D)* to $D - DMOD(D, ONE)$.

```
      SUBROUTINE SLINK(N, N1, N20, DELTA, B, C, DLARGE,
     $  G, H, X, W1, W2, GROUPP, TOPP, PRINTX, IFAULT)
C
C         ALGORITHM AS 15  APPL. STATIST. (1969) VOL.18, P.106
C
C         PERFORMS SINGLE LINKAGE CLUSTERING. INFORMATION IS
C         SUPPLIED BY SUBROUTINE PRTREE, IN ARRAYS B AND C.
C         POINTS ARE LISTED IN SORTED ORDER IN ARRAY G, AND THE
C         CORRESPONDING ARRAY H MARKS THE LAST MEMBER OF EACH
C         CLUSTER WITH A 1, OTHERWISE THE ENTRY IS 0. THE ARRAY
C         X STORES CODE NUMBERS FOR OUTPUT OF THE DENDROGRAM
C
      INTEGER P, Q, R, S, T, U, V, W, B(N), G(N), H(N),
     $  X(N20), W1(N1), W2(N1)
      REAL DELTA, C(N), DLARGE, D, LEVEL, ONE, ZINT
C
      DATA ONE /1.0/
C
      ZINT(D) = AINT(D)
C
      IFAULT = 1
      IF (N1 .NE. N + 1) RETURN
      IF (N20 .NE. 20 * N) RETURN
      IFAULT = 0
C
C         CLUSTERING STARTS AT THE FIRST INTEGRAL MULTIPLE OF
C         DELTA WHICH IS GREATER THAN D, THE SHORTEST LINK OF
C         THE MINIMUM SPANNING TREE
C
      D = C(2)
      IF (N .LT. 3) GOTO 15
      DO 10 I = 3, N
   10 IF (D .GT. C(I)) D = C(I)
   15 DO 20 I = 1, N
      G(I) = I
      H(I) = 1
      X(I) = 3
   20 CONTINUE
      P = 0
      LEVEL = DELTA * (ONE + ZINT(D / DELTA))
C
C         FOR EACH LINK IN ARRAY C THAT IS SHORTER THAN LEVEL,
C         TWO CLUSTERS ARE AMALGAMATED. THE AMALGAMATION
C         INVOLVES A REORDERING OF ARRAYS G AND H AND REMOVAL
C         OF THE END-OF-CLUSTER MARKER FROM THE EARLIER
C         CLUSTER. LINKS ONCE USED ARE INCREASED BY DLARGE TO
C         PREVENT RE-USE
C
```

```
 30 P = P + 1
    DO 150 I = 2, N
    IF (C(I) .GE. LEVEL) GOTO 150
    J = B(I)
    C(I) = C(I) + DLARGE
    K = I
    DO 40 M = 1, N
    IF (G(M) .EQ. J) Q = M
    IF (G(M) .EQ. K) R = M
 40 CONTINUE
    IF (Q .LE. R) GOTO 50
    M = R
    R = Q
    Q = M
 50 DO 60 S = Q, N
    IF (H(S) .NE. 0) GOTO 70
 60 CONTINUE
 70 T = R
 80 T = T - 1
    IF (T .LT. 1) GOTO 90
    IF (H(T) .EQ. 0) GOTO 80
 90 T = T + 1
    H(S) = 0
    L = 0
    DO 100 R = T, N
    L = L + 1
    W1(L) = G(R)
    W2(L) = H(R)
    IF (H(R) .NE. 0) GOTO 110
100 CONTINUE
110 W = S + 1
    M = T
    L = R + 1
120 M = M - 1
    IF (M .LT. W) GOTO 130
    L = L - 1
    G(L) = G(M)
    H(L) = H(M)
    GOTO 120
130 U = R - T + 1
    L = W - 1
    DO 140 M = 1, U
    L = L + 1
    G(L) = W1(M)
    H(L) = W2(M)
140 CONTINUE
150 CONTINUE
    CALL GROUPP(LEVEL, N, G, H)
    W = N * P
    U = 0
    V = 0
    K = 0
C
C       DENDROGRAM INDICATORS ARE NOW COMPILED AND STORED
C       IN X. POINTS ARE EXAMINED IN THE ORDER DEFINED BY
C       THE ARRAY G. S IS THE CORRESPONDING ENTRY ON THE
C       PREVIOUS ITERATION, U = 0 FOR THE FIRST MEMBER OF
C       A CLUSTER, V = 1 WHEN AMALGAMATIONS OCCUR, FROM
C       THE LAST MEMBER OF THE FIRST COMPONENT CLUSTER
C       UNTIL THE LAST MEMBER OF THE AMALGAMATED CLUSTER.
C       K IS THE TOTAL NUMBER OF CLUSTERS
C
    DO 160 I = 2, N
160 K = K + H(I)
    IF (P .GT. 19) GOTO 270
    DO 260 I = 1, N
    J = G(I)
    L = J + W - N
    S = X(L)
```

```
      IF (U .NE. 0) GOTO 190
      IF (H(I) .NE. 1) GOTO 170
      T = 3
      GOTO 250
170   IF (S .NE. 3) GOTO 180
      T = 1
      U = 1
      V = 1
      GOTO 250
180   T = 0
      U = 1
      GOTO 250
190   IF (H(I) .NE. 1) GOTO 210
      IF (V .NE. 0) GOTO 200
      T = 3
      U = 0
      GOTO 250
200   T = 2
      U = 0
      V = 0
      GOTO 250
210   IF (S .LT. 2 .OR. S .GT. 3) GOTO 230
      IF (V .NE. 0) GOTO 220
      T = 1
      U = 1
      V = 1
      GOTO 250
220   T = 5
      U = 1
      GOTO 250
230   IF (V .NE. 0) GOTO 240
      T = 0
      U = 1
      GOTO 250
240   T = 4
250   L = J + W
      X(L) = T
260   CONTINUE
270   LEVEL = LEVEL + DELTA
      IF (K .NE. 1) GOTO 30
      CALL TOPP
      DO 280 I = 1, N
280   CALL PRINTX(I, G, N, P, X, N20)
      RETURN
      END
```

Algorithm AS 30

HALF-NORMAL PLOTTING
By D. N. Sparks
Audits of Great Britain Ltd,
West Gate, London W5 1UA, UK.

Keywords: Half-normal plot; Normality of residuals; Factorial experiments; Analysis of variance.

LANGUAGE

Fortran 66.

DESCRIPTION AND PURPOSE

Given a set of root mean squares obtained from the analysis of a factorial experiment, one method of testing for significance, and/or examining for evidence of certain types of error, is to draw a half-normal plot of these values (Daniel, 1959). The subroutine takes such a set of root mean squares and plots them on a half-normal probability scale via the line printer or other device. For this sort of application the accuracy of the average line printer is more than adequate.

STRUCTURE

SUBROUTINE HNPLOT (IWRITE, IWIDTH, IDEPTH, OBS, N, IOUT, IFAULT)

Formal parameters

IWRITE	Integer	input:	the channel number for output.
IWIDTH	Integer	input:	the width of the plot. Must be between 40 and 200 inclusive. See below for further details.
IDEPTH	Integer	input:	the number of lines of the plot. Must not be less than 15. Including a horizontal axis and scale, the total depth will be *IDEPTH* + 3.
OBS	Real array (*N*)	input:	the values to be plotted, sorted $OBS(1) \leqslant OBS(2) \leqslant \ldots \leqslant OBS(N)$. Values must not be negative, or greater than 999999.
N	Integer	input:	the number of values to be plotted. Must be between 2 and 1250 inclusive.

IOUT	Integer array (*IWIDTH*)	workspace:	
IFAULT	Integer	output:	1 for illegal value of *N*, *IWIDTH* or *IDEPTH*; 2 if *OBS* array not correctly sorted; 3 for illegal *OBS* value; 4 for illegal argument for *PPND* (this fault is probably impossible); 0 otherwise.

Auxiliary algorithm
REAL FUNCTION PPND (P, IFAULT) — Algorithm AS 111 (Beasley and Springer, 1977).

Width of plot
Including a vertical axis and scale the total width will be *IWIDTH* + 12 if the output device is one defined by Fortran as a 'printing' device which takes the first character of the line as a carriage-control character, or *IWIDTH* + 13 otherwise.

RESTRICTIONS

It is assumed that characters, held in 1H form, may be assigned from one integer location to another.

PRECISION

Real precision is perfectly adequate on any computer, so the usual *Applied Statistics* devices to enable easy translation to double precision are not incorporated. If the observations are held in a double precision array, they should be copied to a single precision array to be used as the actual argument corresponding to *OBS*.

EDITORS' REMARKS

Considerable alterations have been made to this algorithm, mainly to make it more flexible for different output devices. When it was first published, to think in terms of line printers only was reasonable although, even then, different widths of printer needed considering. The text said that different widths were 'easily catered for' but the instructions given were only intended to change from one fixed width to another (and were inadequate even for that).

The current version allows successive calls of the algorithm to print on different devices, with different widths and different page depths.

The original included an approximate method of finding normal deviates, which was quite adequate for the purpose, but as Algorithm AS 111 is available it may as well be used instead.

The original, via its first argument, printed a title above the plot. This is omitted. If a title is wanted it can be more easily done as a separate *WRITE* statement before calling *HNPLOT.*

The original took the observations in any order and sorted them (which was incorrect as the text gave the *OBS* array as 'input' only). The current version asks the user to sort the array first, and the algorithm checks that this has been done.

REFERENCES

Beasley, J. D. and Springer, S. G. (1977) Algorithm AS 111. *Appl. Statist.*, **26**, 118–121. (See also this book, page 188).

Daniel, C. (1959) Use of half-normal plots in interpreting factorial two-level experiments, *Technometrics*, **1**, 311–342.

```
      SUBROUTINE HNPLOT(IWRITE, IWIDTH, IDEPTH, OBS, N, IOUT, IFAULT)
C
C        ALGORITHM AS 30  APPL. STATIST. (1970) VOL.19, P.192
C
C        HALF-NORMAL PLOTTING
C
      REAL OBS(N), XPR(9)
      INTEGER DOT, BLANK, PLUS, ZER, IOUT(IWIDTH), IPR(13), IW(10)
      DATA DOT, BLANK, PLUS, ZER, NIN /1H., 1H , 1H+, 1H0, 1H9/
C
C        CHARACTERS AND CONSTANTS FOR PLOTTING X-AXIS.
C
C        THE XPR ARRAY CONTAINS THE FRACTIONS OF THE DISTANCE,
C        FROM THE 0.5 POINT TO THE 0.9999 POINT OF A NORMAL
C        CURVE, OF THE 0.5, 0.6, 0.7, 0.8, 0.9, 0.95, 0.98,
C        0.99 AND 0.999 POINTS
C
      DATA IPR(1), IPR(2), IPR(3), IPR(4), IPR(5), IPR(6)
     $     /1H5,    1H6,    1H7,    1H8,    1H9,    1H9/
      DATA IPR(7), IPR(8), IPR(9), IPR(10), IPR(11), IPR(12), IPR(13)
     $     /1H9,    1H9,    1H9,    1H5,     1H8,     1H9,     1H9/
      DATA XPR(1), XPR(2), XPR(3), XPR(4), XPR(5)
     $      /0.0, 0.0681, 0.1410, 0.2263, 0.3446/
      DATA XPR(6), XPR(7), XPR(8), XPR(9)
     $      /0.4423, 0.5522, 0.6255, 0.8309/
C
    5 FORMAT(F12.4, 200A1)
   10 FORMAT(12X, 200A1)
C
      IFAULT = 1
      IF (N .LT. 2 .OR. N .GT. 1250) RETURN
      IF (IWIDTH .LT. 40 .OR. IWIDTH .GT. 200) RETURN
      IF (IDEPTH .LT. 15) RETURN
      IFAULT = 2
C
C        CHECK THAT VALUES ARE SORTED IN ASCENDING ORDER
C
      DO 20 I = 2, N
      IF (OBS(I) .LT. OBS(I - 1)) RETURN
   20 CONTINUE
      IFAULT = 3
      IF (OBS(1) .LT. 0.0) RETURN
      OBSMAX = OBS(N)
      IF (OBSMAX .GT. 999999.0) RETURN
      IFAULT = 4
```

```
C
C          CALCULATE SCALES FOR THE AXES
C
      XSCALE = FLOAT(IWIDTH) / 3.719
      YSCALE = OBSMAX / FLOAT(IDEPTH)
      XX = 0.5 / FLOAT(N)
      YY = 1.0 - (0.5 * XX)
      L = N
      X = OBSMAX + 0.999 * YSCALE
      M = 1
      DO 160 I = 1, IDEPTH
      X = X - YSCALE
      IF (OBS(L) .LT. X) GOTO 150
      DO 100 IX = 1, IWIDTH
  100 IOUT(IX) = BLANK
C
C          CALCULATE POSITIONING OF POINTS ON PROBABILITY SCALE
C
  110 IX = PPND(YY, IFA) * XSCALE + 1.5
      IF (IFA .GT. 0) RETURN
      YY = YY - XX
      IOUT(IX) = PLUS
      L = L - 1
      IF (L .EQ. 0) GOTO 130
      IF (OBS(L) .GE. X) GOTO 110
C
C          PRINT OUTPUT, LINE BY LINE
C
  130 IF (I .NE. M) GOTO 134
      WRITE (IWRITE, 5) OBSMAX, DOT, (IOUT(IL), IL = 1, IWIDTH)
      IF (I .EQ. 1) GOTO 132
      M = 0
      OBSMAX = 0.0
      GOTO 138
  132 M = IDEPTH / 2 + 1
      OBSMAX = 0.5 * OBSMAX
      GOTO 138
  134 WRITE (IWRITE, 10) DOT, (IOUT(IL), IL = 1, IWIDTH)
  138 IF (L .GT. 0) GOTO 160
      IF (I .EQ. IDEPTH) GOTO 190
      IL = I + 1
      GOTO 170
  150 IF (I .NE. M) GOTO 155
      WRITE (IWRITE, 5) OBSMAX, DOT
      IF (I .EQ. 1) GOTO 152
      M = 0
      OBSMAX = 0.0
      GOTO 160
  152 M = IDEPTH / 2 + 1
      OBSMAX = 0.5 * OBSMAX
      GOTO 160
  155 WRITE (IWRITE, 10) DOT
  160 CONTINUE
      GOTO 190
  170 DO 180 I = IL, IDEPTH
      IF (I .NE. M) GOTO 175
      WRITE (IWRITE, 5) OBSMAX, DOT
      M = 0
      OBSMAX = 0.0
      GOTO 180
  175 WRITE (IWRITE, 10) DOT
  180 CONTINUE
  190 IF (L .LE. 0) GOTO 240
      DO 210 IX = 1, IWIDTH
  210 IOUT(IX) = DOT
  220 IX = PPND(YY, IFA) * XSCALE + 1.5
      IF (IFA .GT. 0) RETURN
      YY = YY - XX
      IOUT(IX) = PLUS
```

```
        L = L - 1
        IF (L .GT. 0) GOTO 220
        WRITE (IWRITE, 5) OBSMAX, (IOUT(IX), IX = 1, IWIDTH)
        GOTO 250
C
C          PRINT PROBABILITY AXIS AND SCALE
C
  240 WRITE (IWRITE, 5) OBSMAX, (DOT, IX = 1, IWIDTH)
  250 DO 260 IX = 1, IWIDTH
  260 IOUT(IX) = BLANK
        X = FLOAT(IWIDTH)
        DO 270 IL = 1, 9
        IX = XPR(IL) * X + 1.5
        IW(IL) = IX
        IOUT(IX) = DOT
  270 CONTINUE
        WRITE (IWRITE, 10) (IOUT(IX), IX = 1, IWIDTH)
        DO 290 IL = 1, 9
        IX = IW(IL)
        IOUT(IX) = IPR(IL)
        IF (IL .LT. 6) IOUT(IX + 1) = ZER
        IF (IL .GT. 5) IOUT(IX + 1) = IPR(IL + 4)
        IF (IL .LT. 9) GOTO 290
        IOUT(IX + 2) = DOT
        IOUT(IX + 3) = NIN
  290 CONTINUE
        WRITE (IWRITE, 10) (IOUT(IX), IX = 1, IWIDTH)
        IFAULT = 0
        RETURN
        END
```

Algorithm AS 41

UPDATING THE SAMPLE MEAN AND DISPERSION MATRIX

By M. R. B. Clarke

Institute of Computer Science

Present address: Queen Mary College, Mile End Road, London, E1 4NS, UK.

Keywords: Mean and dispersion matrix; Sums of squares and products; Multivariate data.

LANGUAGE

Fortran 66 and 77.

DESCRIPTION AND PURPOSE

This algorithm updates the sample mean vector and matrix of corrected sums of squares and products for a set of multivariate data, either when a new unit vector is to be included, or when one is to be withdrawn from the sample. It is particularly useful for exploratory operations such as trying the effect of including and leaving out doubtful observations or outliers.

Method

Let x_i, $i = 1, \ldots, n$ be a set of unit vectors with weights w_i, the w_i depending only on the unit and being the same for each variate. Let

$$W_n = \sum_{i=1}^{n} w_i, \quad \bar{x}_n = \sum_{i=1}^{n} w_i x_i / W_n \text{ and } S_n = \sum_{i=1}^{n} w_i (x_i - \bar{x}_n)(x_i - \bar{x}_n)',$$

where $\bar{x}_1 = x_1$, $S_1 = 0$ and primes denote the transpose. Let $d_{n+1} = x_{n+1} - \bar{x}_n$ and $e_n = x_n - \bar{x}_n$. It can easily be shown by the usual method for partitioning sums of squares that

$$S_{n+1} = S_n + (w_{n+1} - w_{n+1}^2/W_{n+1})d_{n+1}d_{n+1}'$$

and

$$S_{n-1} = S_n - (w_n + w_n^2/W_{n-1})e_n e_n',$$

and correspondingly for the means

$$\bar{x}_{n+1} = \bar{x}_n + w_{n+1}d_{n+1}/W_{n+1}$$

and

$$\bar{x}_{n-1} = \bar{x}_n - w_n e_n/W_{n-1}.$$

If a positive weight corresponds to adding a new vector to the sample and a negative weight to removing one then precisely the same algorithm can be used in both cases.

STRUCTURE

SUBROUTINE DSSP (X, XMEAN, XSSP, WT, SUMWT, NVAR, NN, IFAULT)

Formal parameters

X	Real array (*NVAR*)	input:	unit vector.
		output:	deviation from current mean.
XMEAN	Real array (*NVAR*)	input:	current mean vector.
		output:	updated mean vector.
XSSP	Real array (*NN*)	input:	lower triangle of current matrix of corrected sums of squares and products.
		output:	updated matrix.
WT	Real	input:	weight of *X*.
SUMWT	Real	input:	current sum of weights of vectors in sample. Must be set to 0 at first entry.
		output:	updated sum of weights.
NVAR	Integer	input:	number of variates or components of *X*.
NN	Integer	input:	value of *NVAR* (*NVAR* + 1)/2.
IFAULT	Integer	output:	1 if *SUMWT* < 0 either before or after updating; 2 if *NVAR* < 1 or *NN* ≠ *NVAR* (*NVAR* + 1)/2; 0 otherwise.

Constant

EPS is set to 10^{-4} in a *DATA* statement. If *SUMWT* < *EPS* (before updating) it is assumed that *SUMWT* is intended to be zero, and *SUMWT* and *XSSP* are set to zero, while *XMEAN* takes the value of *X* if *WT* ≥ *EPS* or zero otherwise.

RESTRICTIONS

NVAR, NN, XMEAN, XSSP and *SUMWT* must not be changed between successive calls of a series.

PRECISION

To form a double precision version, change *REAL* to *DOUBLE PRECISION* and give double precision versions of the constants in the *DATA* statement.

It is strongly recommended that double precision be used, except on computers with very accurate single precision representation, not because the algorithm has bad numerical properties, but because many cross-product matrices are nearly singular with correlated variates. Many of the problems that arise with multiple-regression programs can be attributed to inaccurate computation of sums of squares and products.

ACKNOWLEDGEMENTS

I am grateful to a referee for suggesting some improvements in the terminology and coding.

EDITORS' REMARKS

The modifications to this algorithm are more substantial than those to many of the others. The principal change is the omission of *NUNIT* from the argument list. In the original version this formal parameter contained the number of units in the sample, but in particular cases it has been found that *NUNIT* and *SUMWT* could become incompatible with each other. The concept of 'number of units' has, therefore, been abandoned, and the action of the algorithm made to depend on 'sum of weights' only.

```
      SUBROUTINE DSSP(X, XMEAN, XSSP, WT, SUMWT, NVAR, NN, IFAULT)
C
C        ALGORITHM AS 41   APPL. STATIST. (1971) VOL.20, P.206
C
C        UPDATES THE MEAN VECTOR XMEAN, AND THE MATRIX OF
C        CORRECTED SUMS OF SQUARES AND PRODUCTS XSSP,
C        WHEN A DATA VECTOR X WITH WEIGHT WT IS INCLUDED
C        (WT .GT. 0) OR EXCLUDED (WT .LT. 0)
C
      REAL X(NVAR), XMEAN(NVAR), XSSP(NN), WT, SUMWT, B, C, ZERO, EPS
C
      DATA ZERO, EPS /0.0, 1.0E-4/
C
      IFAULT = 2
      IF (NVAR .LT. 1 .OR. NN .NE. NVAR * (NVAR + 1) / 2) RETURN
      IFAULT = 1
      IF (SUMWT .LT. ZERO) RETURN
      IF (SUMWT .GE. EPS) GOTO 20
C
C        SUMWT TAKEN TO BE ZERO SO INITIATE
C
    5 SUMWT = ZERO
      K = 0
      DO 10 I = 1, NVAR
      XMEAN(I) = ZERO
      DO 10 J = 1, I
      K = K + 1
      XSSP(K) = ZERO
   10 CONTINUE
      IF (WT .GE. EPS) GOTO 20
      IFAULT = 0
      RETURN
```

```
C
C
C          UPDATE MEANS AND SUMS OF SQUARES AND PRODUCTS
C
   20 K = 0
      SUMWT = SUMWT + WT
      IF (SUMWT .LT. ZERO) RETURN
      IF (SUMWT .LT. EPS) GOTO 5
      IFAULT = 0
      B = WT / SUMWT
      C = WT - B * WT
      DO 30 I = 1, NVAR
      X(I) = X(I) - XMEAN(I)
      XMEAN(I) = XMEAN(I) + B * X(I)
      DO 30 J = 1, I
      K = K + 1
      XSSP(K) = XSSP(K) + C * X(I) * X(J)
   30 CONTINUE
      RETURN
      END
```

Algorithm AS 45

HISTOGRAM PLOTTING
By D. N. Sparks
Audits of Great Britain Ltd,
West Gate, London, W5 1UA, UK.

Keywords: Histogram plotting.

LANGUAGE

Fortran 66.

DESCRIPTION AND PURPOSE

Given a vector of frequencies, or alternatively the original data, the algorithm
will print a histogram.

STRUCTURE

SUBROUTINE HISTGM (IWRITE, IWIDTH, FREQ, DAT, NN, MM, LENG, IND, IFAULT)

Formal parameters

IWRITE	Integer	input:	the unit number for output.
IWIDTH	Integer	input:	the maximum number of characters to be allowed in each line of the output. Must be at least 26.
FREQ	Integer array (*MM*)	input:	if *IND* is *.FALSE.* the vector of frequencies from which the histogram is printed; otherwise no values needed.
		output:	the actual frequencies used, which may be changed from the input values in some circumstances.
DAT	Real array (*NN*)	input:	if *IND* is *.TRUE.* the original data from which frequencies are to be obtained; otherwise no values needed.
NN	Integer	input:	size of the array *DAT*.

MM	Integer	input:	number of classes into which the data are to be grouped, when given in *DAT,* and size of the array *FREQ.* Must not exceed 28.
LENG	Integer	input:	maximum number of lines of print which the output may occupy. Must be at least 11.
IND	Logical	input:	indicates type of data. *.TRUE.* if the original data are used, *.FALSE.* if a vector of frequencies is supplied.
IFAULT	Integer	output:	1 if *MM* > 28 or *IWIDTH* < 26; 2 if *LENG* < 11; 3 if all *DAT* values are equal; 0 otherwise.

The number of lines available for the printout is given in *LENG.* The actual number of lines of output is determined by the highest frequency. If this is greater than the limit, scaling takes place and a suitable message is printed.

When the original data are supplied, the interval mid-points are printed beneath each column, appropriately scaled when necessary.

PRECISION

It does not seem worth making this algorithm adjustable for double precision. If a double precision array is to correspond to *D,* its values should be copied first to a single precision array to be used instead.

EDITORS' REMARKS

In the original version, *IWRITE* and *IWIDTH* were set in a *DATA* statement, assuming that on any particular computer the same values would always be wanted. They have now been made formal parameters instead to allow, for example, a preliminary look at a histogram on a screen of one width followed by printing on paper of another width.

IND originally served as a fault indicator, as well as indicating the type of input data. These two functions have now been separated, *IND* and *IFAULT* serving the purposes.

New code has been introduced to combine columns as necessary if *MM* is too large compared with *IWIDTH.*

```
      SUBROUTINE HISTGM(IWRITE, IWIDTH, FREQ, DAT, NN, MM,
     $ LENG, IND, IFAULT)
C
C         ALGORITHM AS 45  APPL. STATIST. (1971) VOL.20, P.332
C
C         GIVEN A VECTOR OF FREQUENCIES, OR A VECTOR OF RAW DATA,
C         A HISTOGRAM IS PLOTTED SHOWING THE FREQUENCY DISTRIBUTION
C
      DIMENSION IOUT(28), OUT(28), DAT(NN), FREQ(MM)
      INTEGER FREQ, SCALE
      LOGICAL IND
C
C         DEFINE CHARACTERS FOR PRINTING, AND MAGNITUDE OF
C         SMALLEST ACCEPTABLE NUMBER.
C
      DATA IBLANK, ISTAR, IDASH /1H , 1H*, 1H-/
      DATA ETA /1.0E-38/
C
    1 FORMAT(6H EACH , A1, 8H EQUALS , I4, 7H POINTS /)
    2 FORMAT(1H , I8, 3X, 28(4X, A1))
    3 FORMAT(12H INTERVAL  ), 14(F8.3, 2X))
    4 FORMAT(12H MID-POINTS), 5X, 14(F8.3, 2X))
    5 FORMAT(24H0THE PRINTED VALUES MUST/ 22H BE MULTIPLIED BY 10**, I3)
    6 FORMAT(12H0FREQUENCY  , 28I5)
    7 FORMAT(1H , 120A1)
    8 FORMAT(16H CELLS COMBINED,, I3, 3H AT / 21H A TIME, TO FIT WIDTH)
C
C         CHECK INPUT PARAMETERS AND TYPE OF INPUT
C
      IFAULT = 0
      IF (MM .GT. 28 .OR. IWIDTH .LT. 26) IFAULT = 1
      M = MM
      N = NN
      K = (IWIDTH - 15) / 5
      IF (M .GT. K .AND. IND) M = K
      LENGTH = LENG - 10
      IF (LENGTH .LE. 0) IFAULT = 2
      IF (IFAULT .NE. 0) RETURN
      FM = M
      IF (IND) GOTO 15
      IF (M .LE. K) GOTO 120
      KEY = M / K
      IF (K * KEY .NE. M) KEY = KEY + 1
      L = 1
      DO 10 I = 1, K
      IOUT(I) = 0
      DO 10 J = 1, KEY
      IOUT(I) = IOUT(I) + FREQ(L)
      L = L + 1
      IF (L .GT. M) GOTO 12
   10 CONTINUE
   12 M = I
      DO 14 I = 1, M
   14 FREQ(I) = IOUT(I)
      WRITE (IWRITE, 8) KEY
      GOTO 120
C
C         DEFINE A SUITABLE SCALE
C
   15 DO 20 I = 1, M
   20 FREQ(I) = 0
      XMIN = DAT(1)
      XMAX = XMIN
      DO 30 I = 2, N
      DT = DAT(I)
      IF (DT .LT. XMIN) XMIN = DT
      IF (DT .GT. XMAX) XMAX = DT
   30 CONTINUE
      IF (XMAX - XMIN .GE. ETA * FM) GOTO 35
```

```
        IFAULT = 3
        RETURN
     35 KEY = 1
        KOUNT = 0
     40 R = XMAX - XMIN
        B = XMIN
     50 IF (R .GT. 1.0) GOTO 60
        KOUNT = KOUNT + 1
        R = R * 10.0
        GOTO 50
     60 IF (R .LE. 10.0) GOTO 70
        KOUNT = KOUNT - 1
        R = R / 10.0
        GOTO 60
     70 IF (KEY .GT. 2) GOTO 80
        TK = 10.0 ** KOUNT
        B = B * TK
        IF (B .LT. 0.0 .AND. B .NE. AINT(B)) B = B - 1.0
        B = AINT(B) / TK
        R = (XMAX - B) / FM
        KOUNT = 0
        KEY = KEY + 2
        GOTO 50
     80 STEP = AINT(R)
        IF (STEP .NE. R) STEP = STEP + 1.0
        IF (R .LT. 1.5) STEP = STEP - 0.5
        STEP = STEP / 10.0 ** KOUNT
        IF (KEY .EQ. 4) GOTO 90
        IF (XMAX - XMIN .GT. 0.8 * FM * STEP) GOTO 90
        KOUNT = 1
        KEY = 2
        GOTO 40
     90 XMIN = B
        C = STEP * AINT(B / STEP)
        IF (C .LT. 0.0 .AND. C .NE. B) C = C - STEP
        IF (C + FM * STEP .GE. XMAX) XMIN = C
C
C        CALCULATE FREQUENCIES FOR EACH INTERVAL
C
        DO 110 I = 1, N
        J = (DAT(I) - XMIN) / STEP + 1.0
        FREQ(J) = FREQ(J) + 1
    110 CONTINUE
C
C        PRINT FREQUENCY VECTOR
C
    120 WRITE (IWRITE, 6) (FREQ(I), I = 1, M)
        LINE = M * 5 + 15
        WRITE (IWRITE, 7) (IDASH, I = 1, LINE)
C
C        FIND LARGEST FREQUENCY AND SCALE IF NECESSARY
C
        MAX = 0
        DO 130 I = 1, M
        IF (FREQ(I) .GT. MAX) MAX = FREQ(I)
    130 CONTINUE
        SCALE = 1
        DIV = 1.0
        IF (MAX .LE. LENGTH) GOTO 140
        SCALE = (MAX + LENGTH - 1) / LENGTH
        WRITE (IWRITE, 1) ISTAR, SCALE
        DIV = 1.0 / FLOAT(SCALE)
C
C        CLEAR OUTPUT TO BLANKS
C
    140 DO 150 I = 1, M
    150 IOUT(I) = IBLANK
```

```
C
C           FOR EACH LINE OF PRINT, PLACE OUTPUT CHARACTERS IN
C           THEIR APPROPRIATE POSITIONS IN THE OUTPUT VECTOR
C
        MAX = FLOAT(MAX) * DIV + 0.5
        DO 170 I = 1, MAX
        K = MAX + 1 - I
        DO 160 J = 1, M
        INDEX = FLOAT(FREQ(J)) * DIV + 0.5
        IF (INDEX .EQ. K) IOUT(J) = ISTAR
  160 CONTINUE
        L = K * SCALE
C
C           PRINT LINE OF FREQUENCIES
C
        WRITE (IWRITE, 2) L, (IOUT(J), J = 1, M)
  170 CONTINUE
        WRITE (IWRITE, 7) (IDASH, I = 1, LINE)
        IF (.NOT. IND) RETURN
C
C           COMPUTE INTERVAL MID-POINTS AND SCALE IF NECESSARY
C
        K = 0
        XMIN = XMIN + STEP * 0.5
        XMAX = XMIN + STEP * FLOAT(M - 1)
        XM = AMIN1(ABS(XMIN), ABS(XMAX))
        IF (XM .LT. ETA) XM = XM + STEP
  180 IF (XM .GE. 0.1) GOTO 190
        K = K + 1
        XM = XM * 10.0
        GOTO 180
  190 XM = AMAX1(XMAX, -XMIN)
  200 IF (XM .LT. 1000.0) GOTO 210
        K = K - 1
        XM = XM / 10.0
        GOTO 200
  210 TK = 10.0 ** K
        STEP = STEP * TK
        OUT(1) = XMIN * TK
        DO 220 I = 2, M
  220 OUT(I) = OUT(I - 1) + STEP
C
C           PRINT INTERVAL MID-POINTS
C
        WRITE (IWRITE, 3) (OUT(J), J = 1, M, 2)
        WRITE (IWRITE, 4) (OUT(J), J = 2, M, 2)
        K = -K
        IF (K .NE. 0) WRITE (IWRITE, 5) K
        RETURN
        END
```

Algorithm AS 47

FUNCTION MINIMIZATION USING A SIMPLEX PROCEDURE

By R. O'Neill
University of Bath
Present address: European Ferries Group, Dover, Kent, UK.

Keywords: Function minimization; Optimization; Nelder–Mead procedure.

LANGUAGE

Fortran 66 and 77.

DESCRIPTION AND PURPOSE

The minimum is found of a user-specified function of N variables. The algorithm used is due to Nelder and Mead (1965). On exit, the minimum value of the function is in *YNEWLO*, its co-ordinates in array *XMIN*, and the number of function evaluations performed in *ICOUNT*.

On entry we construct a simplex, i.e. $(N + 1)$ points in N dimensions.

The size, shape and orientation of the simplex are determined by *START(I)* containing the N co-ordinates of the guessed starting point and the vector *STEP(I)*. The values used in *STEP(I)* will depend on the required size of the simplex and the relative magnitudes of the units for each variable. Optimum values of *STEP* and *START* exist but since they imply knowledge of the position of the minimum they obviously cannot be determined. Fortunately, the algorithm will work successfully for all values of *STEP* and *START* but those far from optimum will need more iterations to converge.

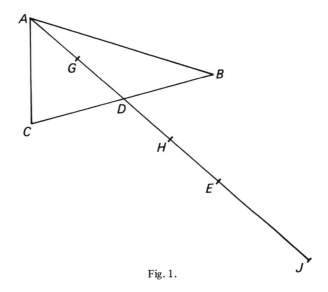

Fig. 1.

The algorithm will work for functions of any number of variables (here limited to 20); our only assumptions about the function are that it is continuous and one-valued.

We illustrate the algorithm with an example of the minimization of a function of two variables. (See Fig. 1 on page 79).

Let $F(A)$ be the function value of the point $A = (X_A, Y_A)$ being a point in the space of the two variables.

Suppose $F(A) >$ both $F(B)$ and $F(C)$ then we replace A by its reflection, E, in D, the centroid of B and C. If E is not successful, i.e. $F(E) \geqslant F(B)$ and $F(C)$, then we contract to G or H depending whether $F(A)$ or $F(E)$ is lower. In the unusual case that $F(G) > F(A)$ (or equivalently $F(H) > F(E)$) then the simplex is contracted around B or C according as $F(B)$ or $F(C)$ is lower.

If, however, E is successful and $F(E) < F(B)$ and $F(C)$ we extend to point J. We then keep E or J as a replacement for A according as $F(E)$ or $F(J)$ is lower.

If neither contraction nor extension is necessary we replace A by E.

Iteration then continues with B, C and the replacement for A, the simplex distorting in shape according to the slopes it encounters.

In the program the 'reflection' (ED/DA), 'extension' (JD/DA) and 'contraction' $(HD/DA = GD/DA)$ coefficients are set at $1 \cdot 0, 2 \cdot 0$ and $0 \cdot 5$ respectively. These are the values recommended by Nelder and Mead as being best for a general situation.

The method for terminating the algorithm is to calculate the variance of the $(N + 1)$ function values occurring at the simplex vertices. If this is less than the user-supplied *REQMIN* then the algorithm is terminated. On input *KCOUNT* contains an upper limit to the number of function evaluations. If this is exceeded the algorithm will be terminated.

To save time we need not make the convergence check at each iteration. The frequency can be controlled using the parameter *KONVGE*.

When the algorithm terminates a check is made to verify whether the point is a local minimum. The value *DEL* is set at $0 \cdot 001 \times STEP(I)$, the original step-length.

The function value is calculated at the suspected minimum $\pm DEL$ along the axis corresponding to each of the N variables.

Only if all $2N$ of these values are greater than the proposed minimum is the latter accepted. If a false minimum is found the simplex is contracted around the lowest point and restarted.

The method is in essence opportunist, only current information being used. However, the simplicity of the algorithm and lack of assumptions about the function or its derivatives make it of very general application.

STRUCTURE

SUBROUTINE NELMIN (FN, N, START, XMIN, YNEWLO, REQMIN, STEP, KONVGE, KCOUNT, ICOUNT, NUMRES, IFAULT)

Formal parameters

FN	Real function		the name of the function to be minimized. See *auxiliary algorithm* below.
N	Integer	input:	the number of variables over which we are minimizing.
START	Real array (*N*)	input:	co-ordinates of the starting point.
		output:	the values may be over-written.
XMIN	Real array (*N*)	output:	co-ordinates of the minimum.
YNEWLO	Real	output:	minimum value of the function.
REQMIN	Real	input:	the terminating limit for the variance of function values.
STEP	Real array (*N*)	input:	determines the size and shape of the initial simplex. The relative magnitudes of its *N* elements should reflect the units of the *N* variables.
KONVGE	Integer	input:	the convergence check is performed every *KONVGE* iterations.
KCOUNT	Integer	input:	maximum number of function evaluations.
ICOUNT	Integer	output:	actual number of function evaluations.
NUMRES	Integer	output:	number of restarts.
IFAULT	Integer	output:	1 if *REQMIN, N* or *KONVGE* has illegal value; 2 if terminated because *KCOUNT* was exceeded without convergence; 0 otherwise.

Constants

The reflection coefficient (*RCOEFF*), extension coefficient (*ECOEFF*) and contraction coefficient (*CCOEFF*) are given the values 1·0, 2·0 and 0·5 in a *DATA* statement. These may be changed if desired.

Auxiliary algorithm
REAL FUNCTION FN (A)

Formal parameters

A   Real array   input: co-ordinates of point
                                                         at which we wish to
                                                         evaluate the function.

The function must assign, to the function name, the function value at the point defined by the elements of A. The function name must be declared as *EXTERNAL* in the routine where *NELMIN* is called.

Example
To minimize the sum of the cubes of two variables the auxiliary algorithm could be

REAL FUNCTION CUBES (A)
REAL A(2)
*CUBES = A(1) ** 3 + A(2) ** 3*
RETURN
END

The calling routine would then use

CALL NELMIN (CUBES, 2, ... etc.).

PRECISION

For a double precision version, change *REAL* to *DOUBLE PRECISION* and give double precision versions of the constants in the *DATA* statement. The auxiliary algorithm must also be modified to a *DOUBLE PRECISION FUNC-TION*, with a *DOUBLE PRECISION* array as its first argument, and to work in double precision internally.

TEST RESULTS

The algorithm was tested on a DEC 2060 computer using four test functions commonly quoted in literature. Each has a function minimum of zero.

(1) Rosenbrock's parabolic valley (Rosenbrock, 1960):
     $y = 100(x_2 - x_1^2)^2 + (1 - x_1)^2$, starting point $(-1 \cdot 2, 1)$.
(2) Powell's quartic function (Powell, 1962):
     $y = (x_1 + 10x_2)^2 + 5(x_3 - x_4)^2 + (x_2 - 2x_3)^4 + 10(x_1 - x_4)^4$, starting point
     $(3, -1, 0, 1)$.

(3) Fletcher and Powell's helical valley (Fletcher and Powell, 1963):

$$y = 100\{x_3 - 10\theta (x_1, x_2)\}^2 + \{\sqrt{(x_1^2 + x_2^2)} - 1\}^2 + x_3^2$$

where $2\pi\theta(x_1, x_2) = \arctan(x_2/x_1)$, $x_1 > 0$

$\qquad\qquad\qquad = \pi + \arctan(x_2/x_1)$, $x_1 < 0$,

starting point $(-1, 0, 0)$.

N.B. Since arctan $(0/0)$ is not defined neither is the function on the line $(0, 0, x_3)$. This line was excluded by assigning a very high value (10,000) to the function there.

(4) $y = \sum_{i=1}^{10} x_i^4$, starting point $(1, 1, \ldots, 1)$.

Results

The parameters were set to be:

$REQMIN = 10^{-16}$: far more accurate than usually necessary.

$STEP(I) = 1 \cdot 0$: side-length of the initial simplex.

$KONVGE = 5$: i.e. convergence tested on every fifth function evaluation.

$KCOUNT = 1,000$: limit on number of function evaluations.

These gave:

Function	Function evaluations performed	No. of restarts	Value of minimum
1	177	0	1.67×10^{-9}
2	266	0	4.03×10^{-9}
3	230	0	2.79×10^{-9}
4	712	14	5.75×10^{-14}

TIME

Execution time is dependent on the number of function evaluations and particularly on the complexity of the function. As an indication of central processor unit time the first two functions (total 443 iterations) took $0 \cdot 23$ seconds while the more involved functions 3 and 4 (total 942 iterations) took $0 \cdot 64$ seconds (DEC 2060).

ACKNOWLEDGEMENTS

I would like to thank Dr J. A. Nelder for his advice and the referees for their useful comments.

This algorithm forms part of a project financed by a grant from the Science Research Council.

EDITORS' REMARKS

The algorithm has been modified to take into account the remarks by Chambers and Ertel (1974), Benyon (1976) and Hill (1976). A corrigendum published in 1974 has also been incorporated, as well as a number of other minor improvements.

The original was in double precision throughout, but in accordance with current *Applied Statistics* policy, this version is single precision with easy change to double if required.

The function *FN* has been put into the argument list to make it easy to minimize more than one function within a single program.

REFERENCES

Benyon, P. R. (1976) Remark AS R15. *Appl. Statist.*, **25**, 97.

Chambers, J. M. and Ertel, J. E. (1974) Remark AS R11. *Appl. Statist.*, **23**, 250–251.

Fletcher, R. and Powell, M. J. D. (1963) A rapidly convergent descent method for minimization. *Computer J.*, **6**, 163–168.

Hill, I. D. (1976) Remark AS R28. *Appl. Statist.*, **27**, 380–382.

Nelder, J. A. and Mead, R. (1965) A simplex method for function minimization. *Computer J.*, **7**, 308–313.

Powell, M. J. D. (1962) An iterative method for finding stationary values of a function of several variables. *Computer J.*, **5**, 147–151.

Rosenbrock, H. (1960) An automatic method for finding the greatest or least value of a function. *Computer J.*, **3**, 175–184.

```
      SUBROUTINE NELMIN(FN, N, START, XMIN, YNEWLO, REQMIN,
     $  STEP, KONVGE, KCOUNT, ICOUNT, NUMRES, IFAULT)
C
C        ALGORITHM AS 47  APPL. STATIST. (1971) VOL.20, P.338
C
C        THE NELDER-MEAD SIMPLEX MINIMIZATION PROCEDURE.
C
C        PURPOSE - TO FIND THE MINIMUM VALUE OF A USER-SPECIFIED
C        FUNCTION.
C
C        AUXILIARY ALGORITHM - THE FUNCTION FN(A) CALCULATES THE
C        FUNCTION VALUE AT POINT A, WHERE A IS AN ARRAY WITH N
C        ELEMENTS.
C
      REAL START(N), XMIN(N), YNEWLO, REQMIN, STEP(N), P(20, 21),
     $  PSTAR(20), P2STAR(20), PBAR(20), Y(21), DN, DNN, Z, YLO,
     $  RCOEFF, YSTAR, ECOEFF, Y2STAR, CCOEFF, RQ, X, DEL, FN, ONE,
     $  HALF, ZERO, EPS
C
      DATA RCOEFF, ECOEFF, CCOEFF, ONE, HALF, ZERO, EPS
     $    / 1.0,    2.0,    0.5, 1.0,  0.5,  0.0, 0.001/
C
C        VALIDITY CHECKS ON INPUT PARAMETERS
C
      IFAULT = 1
      IF (REQMIN .LE. ZERO .OR. N .LT. 1 .OR. N .GT. 20
```

```
     $   .OR. KONVGE .LT. 1) RETURN
       IFAULT = 2
       ICOUNT = 0
       NUMRES = 0
C
       JCOUNT = KONVGE
       DN = N
       NN = N + 1
       DNN = NN
       DEL = ONE
       RQ = REQMIN * DN
C
·C         CONSTRUCTION OF INITIAL SIMPLEX
C
    10 DO 20 I = 1, N
    20 P(I, NN) = START(I)
       Y(NN) = FN(START)
       DO 40 J = 1, N
       X = START(J)
       START(J) = START(J) + STEP(J) * DEL
       DO 30 I = 1, N
    30 P(I, J) = START(I)
       Y(J) = FN(START)
       START(J) = X
    40 CONTINUE
       ICOUNT = ICOUNT + NN
C
C         SIMPLEX CONSTRUCTION COMPLETE
C
C         FIND HIGHEST AND LOWEST Y VALUES. YNEWLO (= Y(IHI))
C         INDICATES THE VERTEX OF THE SIMPLEX TO BE REPLACED
C
    43 YLO = Y(1)
       ILO = 1
       DO 47 I = 2, NN
       IF (Y(I) .GE. YLO) GOTO 47
       YLO = Y(I)
       ILO = I
    47 CONTINUE
    50 YNEWLO = Y(1)
       IHI = 1
       DO 70 I = 2, NN
       IF (Y(I) .LE. YNEWLO) GOTO 70
       YNEWLO = Y(I)
       IHI = I
    70 CONTINUE
C
C         CALCULATE PBAR, THE CENTROID OF THE SIMPLEX
C         VERTICES EXCEPTING THAT WITH Y VALUE YNEWLO
C
       DO 90 I = 1, N
       Z = ZERO
       DO 80 J = 1, NN
    80 Z = Z + P(I, J)
       Z = Z - P(I, IHI)
       PBAR(I) = Z / DN
    90 CONTINUE
C
C         REFLECTION THROUGH THE CENTROID
C
       DO 100 I = 1, N
   100 PSTAR(I) = PBAR(I) + RCOEFF * (PBAR(I) - P(I, IHI))
       YSTAR = FN(PSTAR)
       ICOUNT = ICOUNT + 1
       IF (YSTAR .GE. YLO) GOTO 140
C
C         SUCCESSFUL REFLECTION, SO EXTENSION
C
       DO 110 I = 1, N
```

```
    110 P2STAR(I) = PBAR(I) + ECOEFF * (PSTAR(I) - PBAR(I))
        Y2STAR = FN(P2STAR)
        ICOUNT = ICOUNT + 1
C
C         CHECK EXTENSION
C
        IF (Y2STAR .GE. YSTAR) GOTO 133
C
C         RETAIN EXTENSION
C
        DO 130 I = 1, N
    130 P(I, IHI) = P2STAR(I)
        Y(IHI) = Y2STAR
        GOTO 230
C
C         RETAIN REFLECTION
C
    133 DO 137 I = 1, N
    137 P(I, IHI) = PSTAR(I)
        Y(IHI) = YSTAR
        GOTO 230
C
C         NO EXTENSION
C
    140 L = 0
        DO 150 I = 1, NN
        IF (Y(I) .GT. YSTAR) L = L + 1
    150 CONTINUE
        IF (L .GT. 1) GOTO 133
        IF (L .EQ. 0) GOTO 170
C
C         CONTRACTION ON THE REFLECTION SIDE OF THE CENTROID
C
        DO 160 I = 1, N
    160 P2STAR(I) = PBAR(I) + CCOEFF * (PSTAR(I) - PBAR(I))
        Y2STAR = FN(P2STAR)
        ICOUNT = ICOUNT + 1
        IF (Y2STAR .LE. YSTAR) GOTO 182
C
C         RETAIN REFLECTION
C
        DO 165 I = 1, N
    165 P(I, IHI) = PSTAR(I)
        Y(IHI) = YSTAR
        GOTO 230
C
C         CONTRACTION ON THE Y(IHI) SIDE OF THE CENTROID
C
    170 DO 180 I = 1, N
    180 P2STAR(I) = PBAR(I) + CCOEFF * (P(I, IHI) - PBAR(I))
        Y2STAR = FN(P2STAR)
        ICOUNT = ICOUNT + 1
        IF (Y2STAR .GT. Y(IHI)) GOTO 188
C
C         RETAIN CONTRACTION
C
    182 DO 185 I = 1, N
    185 P(I, IHI) = P2STAR(I)
        Y(IHI) = Y2STAR
        GOTO 230
C
C         CONTRACT WHOLE SIMPLEX
C
    188 DO 200 J = 1, NN
        DO 190 I = 1, N
        P(I, J) = (P(I, J) + P(I, ILO)) * HALF
        XMIN(I) = P(I, J)
    190 CONTINUE
        Y(J) = FN(XMIN)
```

```
  200 CONTINUE
      ICOUNT = ICOUNT + NN
      IF (ICOUNT .GT. KCOUNT) GOTO 260
      GOTO 43
C
C         CHECK IF YLO IMPROVED
C
  230 IF (Y(IHI) .GE. YLO) GOTO 235
      YLO = Y(IHI)
      ILO = IHI
  235 JCOUNT = JCOUNT - 1
      IF (JCOUNT .NE. 0) GOTO 50
C
C         CHECK TO SEE IF MINIMUM REACHED
C
      IF (ICOUNT .GT. KCOUNT) GOTO 260
      JCOUNT = KONVGE
      Z = ZERO
      DO 240 I = 1, NN
  240 Z = Z + Y(I)
      X = Z / DNN
      Z = ZERO
      DO 250 I = 1, NN
  250 Z = Z + (Y(I) - X) ** 2
      IF (Z .GT. RQ) GOTO 50
C
C         FACTORIAL TESTS TO CHECK THAT YNEWLO IS A LOCAL MINIMUM
C
  260 DO 270 I = 1, N
  270 XMIN(I) = P(I, ILO)
      YNEWLO = Y(ILO)
      IF (ICOUNT .GT. KCOUNT) RETURN
      DO 280 I = 1, N
      DEL = STEP(I) * EPS
      XMIN(I) = XMIN(I) + DEL
      Z = FN(XMIN)
      ICOUNT = ICOUNT + 1
      IF (Z .LT. YNEWLO) GOTO 290
      XMIN(I) = XMIN(I) - DEL - DEL
      Z = FN(XMIN)
      ICOUNT = ICOUNT + 1
      IF (Z .LT. YNEWLO) GOTO 290
      XMIN(I) = XMIN(I) + DEL
  280 CONTINUE
      IFAULT = 0
      RETURN
C
C         RESTART PROCEDURE
C
  290 DO 300 I = 1, N
  300 START(I) = XMIN(I)
      DEL = EPS
      NUMRES = NUMRES + 1
      GOTO 10
      END
```

Algorithm AS 51

LOG-LINEAR FIT FOR CONTINGENCY TABLES

By S. J. Haberman

University of Chicago

Present address: Northwestern University, Evanston, Illinois 60201, USA.

Keywords: Maximum likelihood; Log-linear models; Contingency tables; Hierarchical model; Iterative proportional fitting.

LANGUAGE

Fortran 66 and 77.

DESCRIPTION AND PURPOSE

This algorithm performs an iterative proportional fit of the marginal totals of a contingency table. The method used has been described by Deming and Stephan (1940), Fienberg (1970) and Goodman (1970). The algorithm may be used to obtain maximum likelihood estimates which correspond to hierarchical log-linear models for both complete and incomplete contingency tables.

In hierarchical log-linear models for complete tables, the logarithms $\{\mu_{i_1 \ldots i_d}\}$ of the expected values $\{m_{i_1 \ldots i_d}\}$ of the frequencies $\{n_{i_1 \ldots i_d}\}$ of a factorial table with $1 \leqslant i_j \leqslant r_j$ for $j = 1, \ldots, d$ are assumed to satisfy an additive model similar to those encountered in analysis of variance. In such a model,

$$\mu_{i_1 \ldots i_d} = v^\phi + \sum_{k=1}^{l} v^{B_k}_{\{i_j:\ j \in B_k\}}, \tag{1}$$

where the non-empty sets B_k, $k = 1, \ldots, l$, are subsets of the sets of integers from 1 to d. The terms $v^{B_k}_{\{i_j:\ j \in B_k\}}$ satisfy the standard constraints. For example, if in a $2 \times 3 \times 4$ table,

$$\mu_{i_1 i_2 i_3} = v^\phi + v^1_{i_1} + v^2_{i_2} + v^3_{i_3} + v^{12}_{i_1 i_2} + v^{13}_{i_1 i_3} + v^{23}_{i_2 i_3}, \tag{2}$$

then

$$\sum_{i_1=1}^{2} v^1_{i_1} = \sum_{i_2=1}^{3} v^2_{i_2} = \sum_{i_3=1}^{4} v^3_{i_3} = 0$$

and

$$\sum_{i_1=1}^{2} v^{12}_{i_1 i_2} = \sum_{i_2=1}^{3} v^{12}_{i_1 i_2} = \sum_{i_1=1}^{2} v^{13}_{i_1 i_3} = \sum_{i_3=1}^{4} v^{13}_{i_1 i_3}$$

$$= \sum_{i_2=1}^{3} v^{23}_{i_2 i_3} = \sum_{i_3=1}^{4} v^{23}_{i_2 i_3} = 0. \tag{3}$$

In this model, $d = 3$, $B_1 = \{1\}$, $B_2 = \{2\}$, $B_3 = \{3\}$, $B_4 = \{1,2\}$, $B_5 = \{1,3\}$ and $B_6 = \{2,3\}$. In a hierarchical model, it is assumed that if C is a non-empty subset of B_k for some k, $1 \leqslant k \leqslant l$, then $C = B_j$ for some j, $1 \leqslant j \leqslant l$. Thus the example defined by (2) is a hierarchical model.

In any hierarchical model, sets C_k, $k = 1, \ldots, s$, may be found such that each C_k is equal to some $B_{k'}$, each B_k is contained in some $C_{k'}$, and no C_k is contained in any other set $C_{k'}$, $1 \leqslant k' \leqslant s$. The sets $C_1 = \{1,2\}$, $C_2 = \{1,3\}$ and $C_3 = \{2,3\}$ satisfy these requirements in the example of the 2 x 3 x 4 table. If the frequencies in the table are generated from independent Poisson samples or from a multinomial sample, the maximum likelihood estimate $\{\hat{m}_{i_1 \ldots i_d}\}$ of $\{m_{i_1 \ldots i_d}\}$ has the same marginal totals $\hat{m}_{\{i_j : j \in C_k\}}^{C_k}$ as $\{n_{i_1 \ldots i_d}\}$ for $k = 1, \ldots, r$. In the example,

$$\hat{m}_{i_1 i_2}^{12} = \hat{m}_{i_1 i_2 +} = n_{i_1 i_2 +} = n_{i_1 i_2}^{12}, \tag{4}$$

$$\hat{m}_{i_1 i_3}^{13} = \hat{m}_{i_1 + i_3} = n_{i_1 + i_3} = n_{i_1 i_3}^{13} \tag{5}$$

and

$$\hat{m}_{i_2 i_3}^{23} = \hat{m}_{+ i_2 i_3} = n_{+ i_2 i_3} = n_{i_2 i_3}^{23}. \tag{6}$$

Thus a hierarchical model may be described in terms of the marginal totals to be fit.

The Deming–Stephan algorithm is an iterative method which finds $\{\hat{m}_{i_1 \ldots i_d}\}$. The basic iteration cycle begins with an initial estimate $\{m_{i_1 \ldots i_d}^{(0)}\}$ of $\{\hat{m}_{i_1 \ldots i_d}\}$ such that $\{\log m_{i_1 \ldots i_d}^{(0)}\}$ satisfies (1). At the first cycle, one may set each $m_{i_1 \ldots i_d} = 1$ for $1 \leqslant i_j \leqslant r_j$ and $j = 1, \ldots, d$. An iteration cycle consists of s steps. Each step is defined by the equation

$$m_{i_1 \ldots i_d}^{(k)} = m_{i_1 \ldots i_d}^{(k-1)} \frac{n_{\{i_j : j \in C_k\}}^{C_k}}{[m^{(k-1)}]_{\{i_j : j \in C_k\}}^{C_k}}. \tag{7}$$

At the end of a cycle, one may let $m_{i_1 \ldots i_d}^{(0)}$ for the next cycle be $m_{i_1 \ldots i_d}^{(s)}$. In the example,

$$m_{i_1 i_2 i_3}^{(1)} = m_{i_1 i_2 i_3}^{(0)} \frac{n_{i_1 i_2 +}}{m_{i_1 i_2 +}^{(0)}}, \tag{8}$$

$$m_{i_1 i_2 i_3}^{(2)} = m_{i_1 i_2 i_3}^{(1)} \frac{n_{i_1 + i_3}}{m_{i_1 + i_3}^{(1)}} \tag{9}$$

and

$$m_{i_1 i_2 i_3}^{(3)} = m_{i_1 i_2 i_3}^{(2)} \frac{n_{+ i_2 i_3}}{m_{+ i_2 i_3}^{(2)}}. \tag{10}$$

The algorithm may also be applied to incomplete tables in which

$$n_{i_1 i_2 \ldots i_d} = m_{i_1 \ldots i_d} = 0$$

for certain indices and (1) holds for the remaining indices. In such tables, it is sufficient to begin the initial cycle with $m_{i_1\ldots i_d}^{(0)} = 0$ when $m_{i_1\ldots i_d} = 0$ and $m_{i_1\ldots i_d}^{(0)} = 1$ when $m_{i_1\ldots i_d} > 0$. Deming and Stephan (1940) use this algorithm to adjust tables so that they have specified marginal totals.

STRUCTURE

SUBROUTINE LOGLIN (NVAR, DIM, NCON, CONFIG, NTAB, TABLE, FIT, LOCMAR, NMAR, MARG, NU, U, MAXDEV, MAXIT, DEV, NLAST, IFAULT)

Formal parameters

NVAR	Integer	input:	the number of variables d in the table.
DIM	Integer array (*NVAR*)	input:	the number of categories r_j in each variable of the table.
NCON	Integer	input:	the number s of marginal totals to be fit.
CONFIG	Integer array (*NVAR, NCON*)	input:	the sets C_k, $k = 1, \ldots, s$, indicating marginal totals to be fit.
NTAB	Integer	input:	the number of elements in the table.
TABLE	Real array (*NTAB*)	input:	the table to be fit.
FIT	Real array (*NTAB*)	input and output:	the fitted table.
LOCMAR	Integer array (*NCON*)	output:	pointers to the tables in *MARG*.
NMAR	Integer	input:	the dimension of *MARG*.
MARG	Real array (*NMAR*)	output:	the marginal tables to be fit.
NU	Integer	input:	the dimension of *U*.
U	Real array (*NU*)	output:	a work area used to store fitted marginal tables.
MAXDEV	Real	input:	the maximum permissible difference between an observed and fitted marginal total.
MAXIT	Integer	input:	the maximum permissible number of iterations.
DEV	Real array (*MAXIT*)	output:	*DEV(I)* is the maximum observed difference encountered in iteration cycle *I* between an observed and fitted marginal total.

NLAST	Integer	output:	the number of the last iteration.
IFAULT	Integer	output:	an error indicator. See *failure indications* below.

Comments

The tables *TABLE* and *FIT* are arranged in standard Fortran fashion. If the observations to be fit are represented in an array $\{n_{i_1 i_2 i_3}\}$, where $1 \leqslant i_1 \leqslant 2$, $1 \leqslant i_2 \leqslant 3$ and $1 \leqslant i_3 \leqslant 4$, then $NVAR = 3, DIM(1) = 2, DIM(2) = 3, DIM(3) = 4$, $NTAB = 24, TABLE(1) = n_{111}, TABLE(2) = n_{211}, TABLE(3) = n_{121}, TABLE(4) = n_{221}$, etc. If the fitted table is an array $\{\hat{m}_{ijk}\}$, then $FIT(1) = \hat{m}_{111}, FIT(2) = \hat{m}_{211}, FIT(3) = \hat{m}_{121}$, etc.

Marginal totals to be fitted are described by the array *CONFIG*. If C_k has d_k elements, then these elements are given by *CONFIG(J, K)*, where $1 \leqslant J \leqslant d_k$. If $d_k < d$, $CONFIG(J, K) = 0$ for $d_k < J \leqslant d = NVAR$. If maximum likelihood estimates for the table in the preceding paragraph are found under the assumption that (2) holds, then the marginal totals $n_{i_1 i_2 +}$, $n_{i_1 + i_3}$ and $n_{+ i_2 i_3}$ corresponding to $\{1,2\}$, $\{1,3\}$ and $\{2,3\}$ are fitted. Thus $NCON = 3$, $CONFIG(1, 1) = 1$, $CONFIG(2, 1) = 2$, $CONFIG(3, 1) = 0$, $CONFIG(1, 2) = 1$, etc.

FIT is initially set as an estimate of the fitted table. If a complete table is fit to a log-linear model corresponding to the given marginals, one may let $FIT(I) = 1$ for $1 \leqslant I \leqslant NTAB$. If an incomplete table is fit, then one may let $FIT(I) = 1$ if the corresponding fitted value can be positive and $FIT(I) = 0$ if the corresponding fitted value must be zero. If in the three-way table under discussion, n_{111} cannot be observed, one would set $FIT(1) = 0$ and $FIT(I) = 1$ for $2 \leqslant I \leqslant NTAB$.

An iterative procedure is used to transform *FIT* into the desired fitted table. As many as *MAXIT* iterations may be performed, where *MAXIT* may be set as 10 or 15 in most problems. Iteration I is the last iteration if $DEV(I) < MAXDEV$. A reasonable value for *MAXDEV* is often $0 \cdot 25$.

Under some circumstances, convergence occurs after one cycle. In such cases, *MAXIT* may be set to 1 to avoid an unnecessary second cycle. No error condition for failure of convergence will be observed in this case.

The marginals to be fit are placed in *MARG* by the subroutine. They are arranged in the order specified by *CONFIG*. In the example under discussion, *NMAR* must be at least 26. The array *MARG* is constructed so that $MARG(I) = n_{11+}$, $MARG(2) = n_{21+}$, $MARG(3) = n_{12+},\ldots, MARG(7) = n_{1+1}$, $MARG(8) = n_{2+1}$, etc. To facilitate reference to marginal tables, locations of first elements of marginal tables are stored in *LOCMAR*. Thus $LOCMAR(1) = 1, LOCMAR(2) = 7$ and $LOCMAR(3) = 15$.

During computations, a work area U is required. This array should be as large as the largest marginal table used.

Failure indications

If $IFAULT = 0$, then no error was detected.
If $IFAULT = 1$, then *CONFIG* contains an error.

If *IFAULT* = 2, then *MARG* or *U* is too small.

If *IFAULT* = 3, then the algorithm did not converge to the desired accuracy within *MAXIT* iterations.

If *IFAULT* = 4, then *DIM, TABLE, FIT* or *NVAR* is erroneously specified.

RESTRICTIONS

The number of variables *NVAR* must not exceed 7, unless alterations are made as specified in the algorithms.

PRECISION

For a double precision version change *REAL* to *DOUBLE PRECISION* in each of the three subroutines, and give *ZERO* a double precision value in each of the three. In AS 51.2 change *ABS* to *DABS*.

ACKNOWLEDGEMENT

This algorithm was written as part of research sponsored in part by National Science Foundation research grant No. NSF GS 2818.

EDITORS' REMARKS

Previously published corrections have been incorporated and a few minor changes, not affecting functioning, have been made.

REFERENCES

Deming, M. E. and Stephan, F. F. (1940) On a least squares adjustment of a sampled frequency table when the expected marginal totals are known. *Ann. Math. Statist.*, **11**, 427–444.

Fienberg, S. E. (1970) An iterative procedure for estimation in contingency tables. *Ann. Math. Statist.*, **41**, 907–917.

Goodman, L. A. (1970) The multivariate analysis of qualitative data: inter-actions among multiple classifications. *J. Amer. Statist. Ass.*, **65**, 226–256.

```
      SUBROUTINE LOGLIN(NVAR, DIM, NCON, CONFIG, NTAB, TABLE, FIT,
     $ LOCMAR, NMAR, MARG, NU, U, MAXDEV, MAXIT, DEV, NLAST, IFAULT)
C
C        ALGORITHM AS 51   APPL. STATIST. (1972) VOL.21, P.218
C
C        PERFORMS AN ITERATIVE PROPORTIONAL FIT OF THE MARGINAL
C        TOTALS OF A CONTINGENCY TABLE.
C
C        THIS VERSION PERMITS UP TO SEVEN VARIABLES. IF THIS LIMIT
C        IS TOO SMALL, CHANGE THE VALUE OF MAXVAR, AND OF THE
C        DIMENSION IN THE DECLARATION OF CHECK AND ICON - SEE
C        ALSO THE CHANGES NEEDED IN AS 51.1 AND AS 51.2
C
```

```
      REAL MARG(NMAR), U(NU), TABLE(NTAB), FIT(NTAB), MAXDEV,
     $ DEV(MAXIT), ZERO, X, Y, XMAX
      INTEGER DIM(NVAR), CONFIG(NVAR, NCON), LOCMAR(NCON),
     $ POINT, SIZE, ICON(7)
      LOGICAL CHECK(7)
C
      DATA MAXVAR, ZERO /7, 0.0/
C
      IFAULT = 0
      NLAST = 0
C
C        CHECK VALIDITY OF NVAR, THE NUMBER OF VARIABLES,
C        AND OF MAXIT, THE MAXIMUM NUMBER OF ITERATIONS
C
      IF (NVAR .GT. 0 .AND. NVAR .LE. MAXVAR .AND. MAXIT .GT. 0) GOTO 10
    5 IFAULT = 4
      RETURN
C
C        LOOK AT TABLE AND FIT CONSTANTS
C
   10 SIZE = 1
      DO 30 J = 1, NVAR
      IF (DIM(J) .LE. 0) GOTO 5
      SIZE = SIZE * DIM(J)
   30 CONTINUE
      IF (SIZE .LE. NTAB) GOTO 40
   35 IFAULT = 2
      RETURN
   40 X = ZERO
      Y = ZERO
      DO 60 I = 1, SIZE
      IF (TABLE(I) .LT. ZERO .OR. FIT(I) .LT. ZERO) GOTO 5
      X = X + TABLE(I)
      Y = Y + FIT(I)
   60 CONTINUE
C
C        MAKE A PRELIMINARY ADJUSTMENT TO OBTAIN THE FIT
C        TO AN EMPTY CONFIGURATION LIST
C
      IF (Y .EQ. ZERO) GOTO 5
      X = X / Y
      DO 80 I = 1, SIZE
   80 FIT(I) = X * FIT(I)
      IF (NCON .LE. 0 .OR. CONFIG(1, 1) .EQ. 0) RETURN
C
C        ALLOCATE MARGINAL TABLES
C
      POINT = 1
      DO 150 I = 1, NCON
C
C        A ZERO BEGINNING A CONFIGURATION INDICATES
C        THAT THE LIST IS COMPLETED
C
      IF (CONFIG(1, I) .EQ. 0) GOTO 160
C
C        GET MARGINAL TABLE SIZE. WHILE DOING THIS
C        TASK, SEE IF THE CONFIGURATION LIST CONTAINS
C        DUPLICATIONS OR ELEMENTS OUT OF RANGE.
C
      SIZE = 1
      DO 90 J = 1, NVAR
   90 CHECK(J) = .FALSE.
      DO 120 J = 1, NVAR
      K = CONFIG(J, I)
C
C        A ZERO INDICATES THE END OF THE STRING
C
      IF (K .EQ. 0) GOTO 130
C
```

```
C          SEE IF ELEMENT VALID
C
       IF (K .GE. 0 .AND. K .LE. NVAR) GOTO 100
    95 IFAULT = 1
       RETURN
C
C          CHECK FOR DUPLICATION
C
   100 IF (CHECK(K)) GOTO 95
       CHECK(K) = .TRUE.
C
C          GET SIZE
C
       SIZE = SIZE * DIM(K)
   120 CONTINUE
C
C          SINCE U IS USED TO STORE FITTED MARGINALS,
C          SIZE MUST NOT EXCEED NU
C
   130 IF (SIZE .GT. NU) GOTO 35
C
C          LOCMAR POINTS TO MARGINAL TABLES TO BE PLACED IN MARG
C
       LOCMAR(I) = POINT
       POINT = POINT + SIZE
   150 CONTINUE
C
C          GET N, NUMBER OF VALID CONFIGURATIONS
C
       I = NCON + 1
   160 N = I - 1
C
C          SEE IF MARG CAN HOLD ALL MARGINAL TABLES
C
       IF (POINT .GT. NMAR + 1) GOTO 35
C
C          OBTAIN MARGINAL TABLES
C
       DO 190 I = 1, N
       DO 180 J = 1, NVAR
   180 ICON(J) = CONFIG(J, I)
       CALL COLLAP(NVAR, TABLE, MARG, LOCMAR(I), NTAB, NMAR, DIM, ICON)
   190 CONTINUE
C
C          PERFORM ITERATIONS
C
       DO 220 K = 1, MAXIT
C
C          XMAX IS MAXIMUM DEVIATION OBSERVED BETWEEN
C          FITTED AND TRUE MARGINAL DURING A CYCLE
C
       XMAX = ZERO
       DO 210 I = 1, N
       DO 200 J = 1, NVAR
   200 ICON(J) = CONFIG(J, I)
       CALL COLLAP(NVAR, FIT, U, 1, NTAB, NU, DIM, ICON)
       CALL ADJUST(NVAR, FIT, U, MARG, LOCMAR(I), NTAB, NU, NMAR,
      $ DIM, ICON, XMAX)
   210 CONTINUE
C
C          TEST CONVERGENCE
C
       DEV(K) = XMAX
       IF (XMAX .LT. MAXDEV) GOTO 240
   220 CONTINUE
       IF (MAXIT .GT. 1) GOTO 230
       NLAST = 1
       RETURN
```

```
C
C         NO CONVERGENCE
C
  230 IFAULT = 3
      NLAST = MAXIT
      RETURN
C
C         NORMAL TERMINATION
C
  240 NLAST = K
      RETURN
      END
C
      SUBROUTINE COLLAP(NVAR, X, Y, LOCY, NX, NY, DIM, CONFIG)
C
C         ALGORITHM AS 51.1   APPL. STATIST. (1972) VOL.21, P.218
C
C         COMPUTES A MARGINAL TABLE FROM A COMPLETE TABLE.
C         ALL PARAMETERS ARE ASSUMED VALID WITHOUT TEST.
C
C         IF THE VALUE OF NVAR IS TO BE GREATER THAN 7, THE
C         DIMENSIONS IN THE DECLARATIONS OF SIZE AND COORD MUST
C         BE INCREASED TO NVAR+1 AND NVAR RESPECTIVELY.
C
      INTEGER SIZE(8), DIM(NVAR), CONFIG(NVAR), COORD(7)
C
C         THE LARGER TABLE IS X AND THE SMALLER ONE IS Y
C
      REAL X(NX), Y(NY), ZERO
      DATA ZERO /0.0/
C
C         INITIALISE ARRAYS
C
      SIZE(1) = 1
      DO 10 K = 1, NVAR
      L = CONFIG(K)
      IF (L .EQ. 0) GOTO 20
      SIZE(K + 1) = SIZE(K) * DIM(L)
   10 CONTINUE
C
C         FIND NUMBER OF VARIABLES IN CONFIGURATION
C
      K = NVAR + 1
   20 N = K - 1
C
C         INITIALISE Y. FIRST CELL OF MARGINAL TABLE IS
C         AT Y(LOCY) AND TABLE HAS SIZE(K) ELEMENTS
C
      LOCU = LOCY + SIZE(K) - 1
      DO 30 J = LOCY, LOCU
   30 Y(J) = ZERO
C
C         INITIALISE COORDINATES
C
      DO 50 K = 1, NVAR
   50 COORD(K) = 0
C
C         FIND LOCATIONS IN TABLES
C
      I = 1
   60 J = LOCY
      DO 70 K = 1, N
      L = CONFIG(K)
      J = J + COORD(L) * SIZE(K)
   70 CONTINUE
      Y(J) = Y(J) + X(I)
C
C         UPDATE COORDINATES
C
```

```
            I = I + 1
            DO 80 K = 1, NVAR
            COORD(K) = COORD(K) + 1
            IF (COORD(K) .LT. DIM(K)) GOTO 60
            COORD(K) = 0
        80 CONTINUE
            RETURN
            END
C
            SUBROUTINE ADJUST(NVAR, X, Y, Z, LOCZ, NX, NY, NZ, DIM, CONFIG, D)
C
C           ALGORITHM AS 51.2   APPL. STATIST. (1972) VOL.21, P.218
C
C           MAKES PROPORTIONAL ADJUSTMENT CORRESPONDING TO CONFIG.
C           ALL PARAMETERS ARE ASSUMED VALID WITHOUT TEST.
C
C           IF THE VALUE OF NVAR IS TO BE GREATER THAN 7, THE
C           DIMENSIONS IN THE DECLARATIONS OF SIZE AND COORD MUST
C           BE INCREASED TO NVAR+1 AND NVAR RESPECTIVELY.
C
            INTEGER SIZE(8), DIM(NVAR), CONFIG(NVAR), COORD(7)
            REAL X(NX), Y(NY), Z(NZ), D, E, ZERO, ZABS
C
            DATA ZERO /0.0/
C
            ZABS(E) = ABS(E)
C
C           SET SIZE ARRAY
C
            SIZE(1) = 1
            DO 10 K = 1, NVAR
            L = CONFIG(K)
            IF (L .EQ. 0) GOTO 20
            SIZE(K + 1) = SIZE(K) * DIM(L)
        10 CONTINUE
C
C           FIND NUMBER OF VARIABLES IN CONFIGURATION
C
            K = NVAR + 1
        20 N = K - 1
C
C           TEST SIZE OF DEVIATION
C
            L = SIZE(K)
            J = 1
            K = LOCZ
            DO 30 I = 1, L
            E = ZABS(Z(K) - Y(J))
            IF (E .GT. D) D = E
            J = J + 1
            K = K + 1
        30 CONTINUE
C
C           INITIALIZE COORDINATES
C
            DO 40 K = 1, NVAR
        40 COORD(K) = 0
            I = 1
C
C           PERFORM ADJUSTMENT
C
        50 J = 0
            DO 60 K = 1, N
            L = CONFIG(K)
            J = J + COORD(L) * SIZE(K)
        60 CONTINUE
            K = J + LOCZ
            J = J + 1
```

```
C
C         NOTE THAT Y(J) SHOULD BE NON-NEGATIVE
C
      IF (Y(J) .LE. ZERO) X(I) = ZERO
      IF (Y(J) .GT. ZERO) X(I) = X(I) * Z(K) / Y(J)
C
C         UPDATE COORDINATES
C
      I = I + 1
      DO 70 K = 1, NVAR
      COORD(K) = COORD(K) + 1
      IF (COORD(K) .LT. DIM(K)) GOTO 50
      COORD(K) = 0
   70 CONTINUE
      RETURN
      END
```

Algorithm AS 52

CALCULATION OF POWER SUMS OF DEVIATIONS ABOUT THE MEAN

By C. C. Spicer

MRC Computer Unit

Present address: University of Exeter, Exeter, EX4 4PU, UK.

Keywords: Moments; Skewness and kurtosis; Cumulants.

LANGUAGE

Fortran 66 and 77.

DESCRIPTION

If $n - 1$ observations x_i have been read into the computer and a new observation x_n is added then the new power sums are given by:

$$\sum_{i=1}^{n} (x_i - \bar{x}_n)^r = \sum_{i=1}^{n-1} (d_i - \delta x)^r + (n-1)^r (\delta x)^r,$$

where

$\bar{x}_n = $ mean of n observations,

$d_i = (x_i - \bar{x}_{n-1}),$

$\delta x = (x_n - \bar{x}_{n-1})/n,$

whence

$$\sum_{i=1}^{n} (x_i - \bar{x}_n)^2 = \sum_{i=1}^{n-1} (d_i^2) + n(n-1) (\delta x)^2,$$

$$\sum_{i=1}^{n} (x_i - \bar{x}_n)^3 = \sum_{i=1}^{n-1} (d_i^3) - 3(\delta x) \sum_{i=1}^{n-1} (d_i^2) + n(n-1)(n-2)(\delta x)^3,$$

$$\sum_{i=1}^{n} (x_i - \bar{x}_n)^4 = \sum_{i=1}^{n-1} (d_i^4) - 4(\delta x) \sum_{i=1}^{n-1} (d_i^3) + 6(\delta x)^2 \sum_{i=1}^{n-1} (d_i^2)$$

$$+ (n-1) (1 + (n-1)^3) (\delta x)^4.$$

These formulae allow the power sums to be computed with little rounding error but, of course, at the cost of some extra calculation.

STRUCTURE

SUBROUTINE MOMNTS (X, K, N, S1, S2, S3, S4, IFAULT)

Formal parameters

X	Real	input:	value of latest observation.
K	Integer	input:	indicates the power sums required. Must be 1, 2, 3 or 4.
N	Integer	input:	0 at first call, output from previous call subsequently.
		output:	total number of values so far.
$S1$	Real	input:	if $N = 0$ no value needed, otherwise the old mean (output from previous call).
		output:	the new mean.
$S2$	Real	input:	if $N = 0$ or $K = 1$ no value needed, otherwise the value output from previous call.
		output:	new sum of second power (if $K \geqslant 2$).
$S3$	Real	input:	if $N = 0$ or $K < 3$ no value needed, otherwise the value output from previous call.
		output:	new sum of third power (if $K \geqslant 3$).
$S4$	Real	input:	if $N = 0$ or $K < 4$ no value needed, otherwise the value output from previous call.
		output:	new sum of fourth power (if $K = 4$).
$IFAULT$	Integer	output:	1 if value of K or N illegal; 0 otherwise.

RESTRICTIONS

None so far encountered. Serious rounding errors might occur in pathological cases. The values of K, N, $S1$, $S2$, $S3$ and $S4$ must not be changed between calls.

PRECISION

For a double precision version, change *REAL* to *DOUBLE PRECISION*, and give double precision versions of the constants in the *DATA* statement.

EDITORS' REMARKS

The only significant change from the original is that the subroutine is called
MOMNTS instead of *MOMENTS,* to meet Fortran's restriction to six characters
in names.

```
      SUBROUTINE MOMNTS(X, K, N, S1, S2, S3, S4, IFAULT)
C
C         ALGORITHM AS 52   APPL. STATIST. (1972) VOL.21, P.226
C
C         ADDS A NEW VALUE, X, WHEN CALCULATING A MEAN, AND SUMS
C         OF POWERS OF DEVIATIONS. N IS THE CURRENT NUMBER OF
C         OBSERVATIONS, AND MUST BE SET TO ZERO BEFORE FIRST ENTRY
C
      REAL X, S1, S2, S3, S4, AN, AN1, DX, DX2, ZERO, ONE, TWO,
     $   THREE, FOUR, SIX
      DATA ZERO, ONE, TWO, THREE, FOUR, SIX
     $      /0.0, 1.0, 2.0,   3.0, 4.0, 6.0/
C
      IF (K .GT. 0 .AND. K .LT. 5 .AND. N .GE. 0) GOTO 10
      IFAULT = 1
      RETURN
   10 IFAULT = 0
      N = N + 1
      IF (N .GT. 1) GOTO 20
C
C         FIRST ENTRY, SO INITIALISE
C
      S1 = X
      S2 = ZERO
      S3 = ZERO
      S4 = ZERO
      RETURN
C
C         SUBSEQUENT ENTRY, SO UPDATE
C
   20 AN = N
      AN1 = AN - ONE
      DX = (X - S1) / AN
      DX2 = DX * DX
      GOTO (60, 50, 40, 30), K
   30 S4 = S4 - DX * (FOUR * S3 - DX * (SIX * S2 + AN1 *
     $   (ONE + AN1 ** 3) * DX2))
   40 S3 = S3 - DX * (THREE * S2 - AN * AN1 * (AN - TWO) * DX2)
   50 S2 = S2 + AN * AN1 * DX2
   60 S1 = S1 + DX
      RETURN
      END
```

Algorithm AS 57

PRINTING MULTIDIMENSIONAL TABLES

By S. J. Haberman

University of Chicago

Present address: Northwestern University, Evanston, Illinois 60201, USA.

Keywords: Table; Printing; Multidimensional.

LANGUAGE

Fortran 66.

DESCRIPTION AND PURPOSE

This algorithm prints one or more $NVAR$-dimensional parallel tables stored in an array $TABLE$ of length $NTAB$. For example, $TABLE$ might contain a table of observations, a table of fitted values, and a table of residuals. To be specific, suppose that a three-dimensional table $\{n_{ijk}\}$ has been studied, where $1 \leqslant i \leqslant 2$, $1 \leqslant j \leqslant 3$, and $1 \leqslant k \leqslant 4$. From this investigation, a fitted table $\{m_{ijk}\}$ and a residual table $\{r_{ijk}\}$ have been derived. For purposes of display it is desired that the printed table have the format shown in Table 1. In this case, $TABLE$ is divided into three subtables.

Table 1 — Sample test.

Var. 1	Var. 2		Var. 3			
			Cat. 31	Cat. 32	Cat. 33	Cat. 34
Cat. 11	Cat. 21	Obs.	n_{111}	n_{112}	n_{113}	n_{114}
		Fit	m_{111}	m_{112}	m_{113}	m_{114}
		Res.	r_{111}	r_{112}	r_{113}	r_{114}
Cat. 11	Cat. 22	Obs.	n_{121}	n_{122}	n_{123}	n_{124}
		Fit	m_{121}	m_{122}	m_{123}	m_{124}
		Res.	r_{121}	r_{122}	r_{123}	r_{124}
Cat. 11	Cat. 23	Obs.	n_{131}	n_{132}	n_{133}	n_{134}
		Fit	m_{131}	m_{132}	m_{133}	m_{134}
		Res.	r_{131}	r_{132}	r_{133}	r_{134}
Cat. 12	Cat. 21	Obs.	n_{211}	n_{212}	n_{213}	n_{214}
		Fit	m_{211}	m_{212}	m_{213}	m_{214}
		Res.	r_{211}	r_{212}	r_{213}	r_{214}
Cat. 12	Cat. 22	Obs.	n_{221}	n_{222}	n_{223}	n_{224}
		Fit	m_{221}	m_{222}	m_{223}	m_{224}
		Res.	r_{221}	r_{222}	r_{223}	r_{224}
Cat. 12	Cat. 23	Obs.	n_{231}	n_{232}	n_{233}	n_{234}
		Fit	m_{231}	m_{232}	m_{233}	m_{234}
		Res.	r_{231}	r_{232}	r_{233}	r_{234}

The subtable corresponding to the observations $\{n_{ijk}\}$ begins at $TABLE(1)$, the subtable for the fit $\{m_{ijk}\}$ begins at $TABLE(25)$, and the subtable of residuals $\{r_{ijk}\}$ begins at $TABLE(49)$. Each subtable is arranged in standard Fortran fashion; that is, $TABLE(1) = n_{111}$, $TABLE(2) = n_{211}$, $TABLE(3) = n_{121}$, etc.

To permit proper printing of the table, the algorithm requires descriptive information concerning the structure of the table, and information which indicates how the table should be displayed. The basic structural information is provided by an integer array DIM of length $NVAR$ and an integer array LOC of length COL. The array DIM gives the number of categories in each of the table variables. Thus $DIM(1) = 2$, $DIM(2) = 3$ and $DIM(3) = 4$ in the example. The array LOC gives the location in $TABLE$ of each of the COL parallel tables. In the example, $COL = 3$, $LOC(1) = 1$, $LOC(2) = 25$ and $LOC(3) = 49$.

Label information is required for the table variables, the categories of each variable and the names of the parallel tables. In addition, the title for the complete display is needed. The real arrays $TITLE$, $VARNAM$, $CATNAM$ and $COLLAB$ provide these data. Each label of $VARNAM$, $CATNAM$ and $COLLAB$ consists of $WORDS$ machine words. The title consists of $NT2$ lines with $NT1$ words per line. Of the $NVAR$ labels of $VARNAM$, the last $VERT$ of them are printed on the left-hand side of the page, while $NVAR-VERT$ labels are printed along the top. In the example, $VERT = 2$.

The body of the table is printed by use of F-conversion. The format for an entry from parallel table J is $Fa.b$, where a is $WORDS \times LENGTH + 4$ and b is $DEC(J)$. Here $LENGTH$ is the number of characters stored in a machine word. It is specified in a $DATA$ statement, which is used to define the machine configuration in use. The integer array DEC has length COL.

When many tables are to be printed, it is sometimes useful to print several small tables on the same page. To permit this practice, the user may employ the line counter $LINE$ and the carriage control word $RESTOR$. If $RESTOR$ is $.FALSE.$ and the table to be printed can fit within the remainder of the page, then no carriage restore is made before the table is printed. The line counter $LINE$ indicates the location on the page of the last line of the table.

STRUCTURE

SUBROUTINE TABWRT (TITLE, NT1, NT2, TABLE, NTAB, DIM, NVAR, LOC, COL, DEC, VARNAM, WORDS, CATNAM, MAXCAT, COLLAB, VERT, RESTOR, LINE, SKIP, PAGE, WIDTH, UNIT, IFAULT)

Formal parameters

TITLE	Real array($NT1$, $NT2$)	input:	the title of the table.
NT1	Integer	input:	the number of words per line in the title.
NT2	Integer	input:	the number of lines in the title.

TABLE	Real array(*NTAB*)	input:	the tables to be printed.
NTAB	Integer	input:	the number of elements in *TABLE*.
DIM	Integer array(*NVAR*)	input:	the number of categories in each variable of the table.
NVAR	Integer	input:	the number of variables (dimensions) in the table.
LOC	Integer array(*COL*)	input:	the locations in *TABLE* of the subtables to be printed.
COL	Integer	input:	the number of subtables to be printed.
DEC	Integer array(*COL*)	input:	the number of places to the right of the decimal point for each subtable.
VARNAM	Real array(*WORDS, NVAR*)	input:	the variable names.
WORDS	Integer	input:	the number of machine words per name.
CATNAM	Real array(*WORDS, MAXCAT, NVAR*)	input:	the names of the variable categories.
MAXCAT	Integer	input:	the maximum number of categories per variable.
COLLAB	Real array(*WORDS, COL*)	input:	the table names.
VERT	Integer	input:	the number of labels to be printed on the left side of the page.
RESTOR	Logical	input:	the carriage is always restored before the table is printed if *RESTOR* is *TRUE*.
LINE	Integer	input: and output:	the line counter.
SKIP	Integer	input:	the number of lines between tables on the same page.
PAGE	Integer	input:	the number of lines per page.
WIDTH	Integer	input:	the number of characters per line (excluding carriage-control character).
UNIT	Integer	input:	the number designating the output device.
IFAULT	Integer	output:	1 if *NVAR* or *DIM* is incorrectly specified; 2 if the table is too large; 0 otherwise.

Auxiliary algorithm
FUNCTION CONVRT(I)

CONVRT changes an integer I into its alphanumeric equivalent. The version of *CONVRT* given as AS 57.1 is suitable for universal use with Fortran 66 compilers. More elegant versions of this routine can be constructed which are machine-dependent.

Adjustable constants
The following values are defined in a *DATA* statement:

LENGTH	Integer	data:	the number of characters per integer word location.
MAXVAR	Integer	data:	the maximum number of variables permitted in a table.

The *DATA* statement listed in the program is suitable for an IBM 360 computer. It should be noted that the dimension of *SIZE* should be *MAXVAR* +1 and the dimension of *COORD* should be *MAXVAR*.

RESTRICTIONS

The compiler must permit assignment statements in which the right-hand side is a variable containing an H (alphanumeric) value, and function values which are H values must also be acceptable.

ACKNOWLEDGEMENTS

The author wishes to acknowledge Mr Shang-Ping Lin's assistance in testing this subroutine. Research for this algorithm was supported in part by National Science Foundation Grant No. GS31967X and in part by the Department of Statistics, University of Chicago.

PRECISION

It does not seem worth making this algorithm adjustable for double precision. If a double precision array is to correspond to *TABLE*, its values should be copied first to a single precision array to be used instead.

EDITORS' REMARKS

In the original version, *SKIP, PAGE, WIDTH* and *UNIT* were internal constants, with values set in a *DATA* statement. Here they have, instead, been made additional formal parameters. This allows much greater flexibility, for example in examining possible outputs on a screen before changing the *UNIT* number to print on paper.

The algorithm is not compatible with Fortran 77, because of the different character-handling in that language.

```
      SUBROUTINE TABWRT(TITLE, NT1, NT2, TABLE, NTAB, DIM, NVAR,
     $ LOC, COL, DEC, VARNAM, WORDS, CATNAM, MAXCAT, COLLAB,
     $ VERT, RESTOR, LINE, SKIP, PAGE, WIDTH, UNIT, IFAULT)
C
C         ALGORITHM AS 57  APPL. STATIST. (1973) VOL.22, P.118
C
C         TABWRT PRINTS MULTIDIMENSIONAL TABLES.
C
C         A TABLE CAN HAVE AT MOST MAXVAR VARIABLES. NOTE THAT
C         SIZE HAS DIMENSION MAXVAR+1 AND COORD HAS DIMENSION MAXVAR
C
      LOGICAL RESTOR
      INTEGER DIM(NVAR), COL, LOC(COL), DEC(COL), WORDS, VERT
      INTEGER SKIP, PAGE, WIDTH, UNIT
      INTEGER SIZE(8), COORD(7), HOR, DOWN, SPACE, ONEPAG
C
C         ARRAYS CONTAINING DATA AND LABELS
C
      REAL TITLE(NT1, NT2), TABLE(NTAB), VARNAM(WORDS, NVAR),
     $ CATNAM(WORDS, MAXCAT, NVAR), COLLAB(WORDS, COL)
      REAL OUT(33), CHAR(3)
      REAL TAB1, TAB2, XLAB1, XLAB2, XLAB3
C
C         VARIABLE FORMATS.
C
      REAL FMT(15), FMTT(7)
      DATA FMTT(1),FMTT(2),FMTT(3),FMTT(4),FMTT(5),FMTT(6),FMTT(7)
     $   /4H(A1, ,1H       ,2HX,   ,1H       ,1HA  ,1H       ,1H)   /
C
      DATA FMT(1),FMT(2),FMT(3),FMT(4),FMT(5),FMT(6),FMT(7),FMT(8)
     $   /4H(A1,,1H       ,2HX,   ,1H       ,4H(4X,,1H       ,1HA  ,1H   /
C
      DATA FMT(9),FMT(10),FMT(11),FMT(12),FMT(13),FMT(14),FMT(15)
     $   /2H),   ,1H       ,1HF   ,1H       ,1H.   ,1HO    ,1H)   /
C
C         CARRIAGE CONTROL.
C
      DATA CHAR(1), CHAR(2), CHAR(3) /1H , 1HO, 1H1/
C
C         ADJUSTABLE CONSTANTS - SEE INTRODUCTORY TEXT
C
      DATA LENGTH, MAXVAR /4, 7/
C
C         SET ERROR INDICATOR AND GET SIZES.
C
      IFAULT = 0
      IF (NVAR .GT. MAXVAR) GOTO 98
      SIZE(1) = 1
      DO 1 I = 1, NVAR
      COORD(I) = 1
      IF (DIM(I) .LE. 0) GOTO 2
      IF (DIM(I) .GT. MAXCAT) GOTO 98
      SIZE(I + 1) = SIZE(I) * DIM(I)
    1 CONTINUE
      I = NVAR + 1
    2 N = I - 1
      DOWN = VERT
```

```
C
C           ALLOCATE VARIABLES TO HORIZONTAL AND VERTICAL SCALES.
C
      IF (DOWN .GT. N) DOWN = N
      IF (DOWN .LT. 1) DOWN = 1
      HOR = N - DOWN
C
C           LABELW IS WIDTH OF LABEL.
C
      LABELW = WORDS * LENGTH + 4
      I = HOR + 1
      K = I
      DO 3 J = 1, I
C
C           GET LINE WIDTH.
C
      LINEW = LABELW * (SIZE(K) + DOWN + 1)
      IF (LINEW .LT. WIDTH) GOTO 4
      K = K - 1
      DOWN = DOWN + 1
    3 CONTINUE
C
C           NOTE THAT LINE IS NOT WIDE ENOUGH.
C
      IFAULT = 2
      RETURN
    4 HOR = N - DOWN
C
C           SET FORMATS. CONVRT IS AN AUXILIARY
C           INTEGER TO ALPHANUMERIC ROUTINE.
C
      FMTT(2) = CONVRT((WIDTH - LENGTH * NT1) / 2 + 1)
      FMTT(4) = CONVRT(NT1)
      FMTT(6) = CONVRT(LENGTH)
      FMT(8) = FMTT(6)
      I = (WIDTH - LINEW) / 2 + 1
      TAB1 = CONVRT(I)
      TAB2 = CONVRT(I + LABELW * DOWN)
      XLAB1 = CONVRT(DOWN + 1)
      XLAB2 = CONVRT(1)
      FMT(6) = CONVRT(WORDS)
      FMT(12) = CONVRT(LABELW)
C
C           NOW ASCERTAIN VERTICAL SPACE REQUIREMENTS.
C
      SPACE = (COL + 1) * SIZE(N + 1) / SIZE(HOR + 1)
      ONEPAG = SPACE + 2 + HOR + NT2 + NT2 + SKIP
      IF (RESTOR .OR. LINE + ONEPAG .GT. PAGE) LINE = 0
      ICAR = 2
C
C           SEE IF CARRIAGE RESTORE NEEDED
C
      IF (LINE .NE. 0) GOTO 5
      ICAR = 3
      GOTO 7
C
C           SKIP APPROPRIATE NUMBER OF LINES.
C
    5 DO 6 I = 1, SKIP
    6 WRITE (UNIT, 100)
  100 FORMAT(1H )
    7 I = LINE + ONEPAG
      IF (I .LE. PAGE) LINE = I
C
C           SEE HOW MUCH FITS ON A PAGE.
C
      INDEX = N + 1
      K = PAGE - 2 - HOR - NT2 - NT2
      DO 8 I = 1, DOWN
```

```
        IF (SPACE .LE. K) GOTO 9
        INDEX = INDEX - 1
        SPACE = SPACE / DIM(INDEX)
      8 CONTINUE
C
C         RETURN WITH IFAULT = 2 IF TOO LITTLE SPACE AVAILABLE.
C
        IFAULT = 2
        RETURN
C
C         NOTCH IS UNITS PER PAGE.
C
      9 NOTCH = K / SPACE * SIZE(INDEX)
        INC = SIZE(HOR + 1)
        INC1 = INC - 1
        XLAB3 = CONVRT(INC + 1)
        FMT(10) = CONVRT(INC)
        NUM = (SIZE(N + 1) + NOTCH - 1) / NOTCH * NOTCH
        LL = HOR + 1
C
C         SET MARKER.
C
        MARK = 0
        INDEX = WORDS * DOWN
C
C         PRINT A PAGE.
C
        DO 18 I = NOTCH, NUM, NOTCH
        IC = ICAR
C
C         PRINT TITLE.
C
        DO 10 K = 1, NT2
        WRITE (UNIT, FMTT) CHAR(IC), (TITLE(J, K), J = 1, NT1)
        IC = 1
     10 CONTINUE
C
C         SKIP TWO LINES.
C
        WRITE (UNIT, 101)
    101 FORMAT(1H0)
C
C         PRINT HORIZONTAL LABELS.
C
        IF (HOR .EQ. 0) GOTO 12
        FMT(2) = TAB2
        FMT(4) = XLAB3
        DO 11 K = 1, HOR
        L = HOR - K + 1
        I1 = SIZE(L)
        I2 = SIZE(L + 1)
        I3 = DIM(L)
        WRITE (UNIT, FMT) CHAR(1), (VARNAM(I4, L), I4 = 1, WORDS),
      $   ((((CATNAM(I4, I5, L), I4 = 1, WORDS), I6 = 1, I1),
      $   I5 = 1, I3), I7 = I2, INC, I2)
     11 CONTINUE
C
C         VERTICAL LABELS.
C
     12 J = 0
        FMT(2) = TAB1
        FMT(4) = XLAB1
        M = N + 1
        DO 13 K = LL, N
        M = M - 1
        DO 13 I1 = 1, WORDS
        J = J + 1
        OUT(J) = VARNAM(I1, M)
     13 CONTINUE
        WRITE (UNIT, FMT) CHAR(1), (OUT(J), J = 1, INDEX)
```

```
C
C          NOW PRINT BODY OF TABLE.
C
      DO 18 I2 = INC, NOTCH, INC
      J = 0
      M = N + 1
      DO 14 K = LL, N
      M = M - 1
      I3 = COORD(M)
      DO 14 I1 = 1, WORDS
      J = J + 1
      OUT(J) = CATNAM(I1, I3, M)
   14 CONTINUE
      I3 = LOC(1) + MARK
      I4 = I3 + INC1
      FMT(2) = TAB1
      FMT(4) = XLAB1
      FMT(14) = CONVRT(DEC(1))
      WRITE (UNIT, FMT) CHAR(2), (OUT(J), J = 1, INDEX),
     $ (COLLAB(J, 1), J = 1, WORDS), (TABLE(I5), I5 = I3, I4)
      IF (COL .LE. 1) GOTO 16
      FMT(2) = TAB2
      FMT(4) = XLAB2
      DO 15 I6 = 2, COL
      I3 = LOC(I6) + MARK
      I4 = I3 + INC1
      FMT(14) = CONVRT(DEC(I6))
      WRITE (UNIT, FMT) CHAR(1), (COLLAB(I5, I6), I5 = 1, WORDS),
     $ (TABLE(I5), I5 = I3, I4)
   15 CONTINUE
   16 MARK = MARK + INC
C
C          RESET COORDINATES.
C
      DO 17 I1 = LL, N
      COORD(I1) = COORD(I1) + 1
      IF (COORD(I1) .LE. DIM(I1)) GOTO 18
      COORD(I1) = 1
   17 CONTINUE
      RETURN
   18 CONTINUE
C
C          ERROR RETURN.
C
   98 IFAULT = 1
      RETURN
      END
C
      FUNCTION CONVRT(I)
C
C          ALGORITHM AS 57.1  APPL. STATIST. (1973) VOL.22, P.118
C
C          CONVERT FROM INTEGER TO ALPHANUMERIC.
C          THE NORMAL RANGE FOR CONVRT IS FROM 0 TO 101.
C          ALL INTEGERS TO BE CONVERTED WILL BE WITHIN THIS
C          RANGE IF WIDTH DOES NOT EXCEED 131, LENGTH IS AT
C          LEAST 4, AND MAXVAR DOES NOT EXCEED 7.
C
      REAL C(102)
      DATA  C(1), C(2), C(3), C(4), C(5) / 1H0, 1H1, 1H2, 1H3, 1H4/
      DATA  C(6), C(7), C(8), C(9),C(10) / 1H5, 1H6, 1H7, 1H8, 1H9/
      DATA C(11),C(12),C(13),C(14),C(15) /2H10,2H11,2H12,2H13,2H14/
      DATA C(16),C(17),C(18),C(19),C(20) /2H15,2H16,2H17,2H18,2H19/
      DATA C(21),C(22),C(23),C(24),C(25) /2H20,2H21,2H22,2H23,2H24/
      DATA C(26),C(27),C(28),C(29),C(30) /2H25,2H26,2H27,2H28,2H29/
      DATA C(31),C(32),C(33),C(34),C(35) /2H30,2H31,2H32,2H33,2H34/
      DATA C(36),C(37),C(38),C(39),C(40) /2H35,2H36,2H37,2H38,2H39/
      DATA C(41),C(42),C(43),C(44),C(45) /2H40,2H41,2H42,2H43,2H44/
      DATA C(46),C(47),C(48),C(49),C(50) /2H45,2H46,2H47,2H48,2H49/
```

```
DATA C(51),C(52),C(53),C(54),C(55)  /2H50,2H51,2H52,2H53,2H54/
DATA C(56),C(57),C(58),C(59),C(60)  /2H55,2H56,2H57,2H58,2H59/
DATA C(61),C(62),C(63),C(64),C(65)  /2H60,2H61,2H62,2H63,2H64/
DATA C(66),C(67),C(68),C(69),C(70)  /2H65,2H66,2H67,2H68,2H69/
DATA C(71),C(72),C(73),C(74),C(75)  /2H70,2H71,2H72,2H73,2H74/
DATA C(76),C(77),C(78),C(79),C(80)  /2H75,2H76,2H77,2H78,2H79/
DATA C(81),C(82),C(83),C(84),C(85)  /2H80,2H81,2H82,2H83,2H84/
DATA C(86),C(87),C(88),C(89),C(90)  /2H85,2H86,2H87,2H88,2H89/
DATA C(91),C(92),C(93),C(94),C(95)  /2H90,2H91,2H92,2H93,2H94/
DATA C(96),C(97),C(98),C(99),C(100)/2H95,2H96,2H97,2H98,2H99/
DATA C(101),C(102)                       /3H100,3H101/
J = I + 1
IF (J .LT. 1) J = 1
IF (J .GT. 102) J = 102
CONVRT = C(J)
RETURN
END
```

Algorithm AS 60

LATENT ROOTS AND VECTORS OF A SYMMETRIC MATRIX

By D. N. Sparks   and   A. D. Todd

Audits of Great Britain Ltd.   *Rothamsted Experimental Station,*
West Gate, London, W5 1UA, UK   *Harpenden, Herts, AL5 2JQ, UK*

Keywords: Latent roots and vectors; Eigenvalues; QL technique; Tridiagonalization.

LANGUAGE

Fortran 66 and 77.

DESCRIPTION AND PURPOSE

The subroutine *TDIAG* reduces a real symmetric matrix stored in lower triangular form to tridiagonal form, using Householder's reduction. The subroutine *LRVT* finds the latent roots and vectors of a symmetric tridiagonal matrix, given its diagonal elements and subdiagonal elements in two arrays, using *QL* transformations.

The subroutine *TDIAG* is based on the Algol procedure *TRED2* given in Martin *et al.* (1968), and the method is described there and more fully by Wilkinson (1965). The subroutine *LRVT* is based on the Algol procedure *TQL2* given by Bowdler *et al.* (1968), who describe the method. These Algol procedures have now been published in book form — Wilkinson and Reinsch (1971).

STRUCTURE

SUBROUTINE TDIAG (N, TOL, A, D, E, Z, IFAULT)

Formal parameters

N	Integer	input:	order of the real symmetric matrix *A*.
TOL	Real	input:	machine-dependent constant. Set equal to *eta/ precis* where *eta* = the smallest positive number representable in the machine, and *precis* = the smallest positive number for which 1 + *precis* ≠ 1.
A	Real array (*N, N*)	input:	the elements of the symmetric matrix. Only the lower triangle is used.
D	Real array (*N*)	output:	the diagonal elements of the tridiagonal matrix.

E	Real array (N)	output:	$E(2)$ to $E(N)$ give the $N-1$ sub-diagonal elements of the tridiagonal matrix. $E(1)$ is set to zero.
Z	Real array (N, N)	output:	the product of the Householder transformation matrices.
IFAULT	Integer	output:	a fault indicator, equal to 1 if $N \leqslant 1$, and 0 otherwise.

SUBROUTINE LRVT (N, PRECIS, D, E, Z, IFAULT)

Formal parameters

N	Integer	input:	order of the tridiagonal matrix.
PRECIS	Real	input:	the smallest positive number representable on the computer for which $1 + PRECIS \neq 1$.
D	Real array (N)	input:	diagonal elements of the tridiagonal matrix.
		output:	the latent roots.
E	Real array (N)	input:	subdiagonal elements of the tridiagonal matrix in elements $E(2)$ to $E(N)$.
		output:	the values in E are overwritten.
Z	Real array (N, N)	input:	equal to identity if the latent vectors of the tridiagonal matrix are required; or to the Householder transformation output from *TDIAG* if the latent vectors of the full matrix are required.
		output:	the normalized latent vectors, column by column.
IFAULT	Integer	output:	a fault indicator, equal to 1 if more than *MITS* iterations are required, 2 if $N \leqslant 1$, and 0 otherwise. *MITS* is set to 30 in a *DATA* statement, but may be reset by the user, if desired.

RELATED ALGORITHMS

In Martin *et al.* (1968), several other Algol procedures are given for Householder's reduction.

TRED1: Used when latent roots only, or certain selected latent roots and the corresponding vectors are required.

TRED3: As *TRED1*, but with a more economical storage arrangement and marginally slower.

TRED4: Acting as *TDIAG* or *TRED2*, but with the storage arrangement of *TRED3*.

TRBAK1: Derives latent vectors of the original matrix corresponding to those of the transformed matrix using details output from *TRED1*.

TRBAK3: As *TRBAK1*, but in conjunction with *TRED3*.

In Bowdler *et al.* (1968) an Algol procedure *TQL1* is given, for use when only the latent roots are required. In addition, reference is made to other procedures which can be used for solving the generalized eigen problem $\mathbf{AX} = \mathbf{BX}$, $\mathbf{ABX} = \mathbf{X}$, etc.

ACCURACY

The same test matrices were used for these Fortran subroutines as for the original Algol procedures, with satisfactory results. For details of these and further tests involving this algorithm see Sparks and Todd (1973).

RESTRICTIONS

None. Some compilers will allow A to coincide with Z in *TDIAG*, and the algorithm is so written that no harm will be done, but the input matrix will be lost; calling the routine in this way is illegal in standard Fortran.

PRECISION

The version of Algorithm AS 60 given below is in single precision. To obtain a double precision version the following changes should be made:

(1) Change *REAL* to *DOUBLE PRECISION* in both subroutines.
(2) Change the constants in the *DATA* statements to double precision versions.
(3) Change *SQRT* to *DSQRT* in the statement functions in both routines, and *ABS* to *DABS* in *LRVT*.

EDITORS' REMARKS

This algorithm has been modified to incorporate the Corrigendum of 1974, Issue 1. The routines are protected against being called with invalid values of N, necessitating the introduction of an additional parameter *IFAULT* to *TDIAG*.

REFERENCES

Bowdler, H., Martin, R. S., Reinsch, C. and Wilkinson, J. H. (1968) The *QR* and *QL* algorithms for symmetric matrices. *Numer. Math.*, **11**, 293–306.

Martin, R. S., Reinsch, C. and Wilkinson, J. H. (1968) Householder's tri-diagonalisation of a symmetric matrix. *Numer. Math.*, **11**, 181–195.

Martin, R. S. and Wilkinson, J. H. (1968) The implicit *QL* algorithm. *Numer. Math.*, **12**, 377–383.

Sparks, D. N. and Todd, A. D. (1973) A comparison of Fortran subroutines for calculating latent roots and vectors. *Appl. Statist.*, **22**, 220–225.

Wilkinson, J. H. (1965) *The Algebraic Eigenvalue Problem.* London: Oxford University Press.

Wilkinson, J. H. and Reinsch, C. (1971) *Handbook for Automatic Computation,* Vol. 2. *Linear Algebra.* Wien — New York: Springer-Verlag.

```
      SUBROUTINE TDIAG(N, TOL, A, D, E, Z, IFAULT)
C
C         ALGORITHM AS 60.1  APPL. STATIST. (1973) VOL.22, P.260
C
C         REDUCES REAL SYMMETRIC MATRIX TO TRIDIAGONAL FORM
C
      REAL A(N, N), D(N), E(N), F, G, H, HH, TOL, Z(N, N), ZERO,
     $ ONE, ZSQRT
C
      DATA ZERO, ONE /0.0, 1.0/
C
      ZSQRT(H) = SQRT(H)
C
      IFAULT = 1
      IF (N .LE. 1) RETURN
      IFAULT = 0
      DO 10 I = 1, N
      DO 10 J = 1, I
   10 Z(I, J) = A(I, J)
      I = N
      DO 70 I1 = 2, N
      L = I - 2
      F = Z(I, I - 1)
      G = ZERO
      IF (L .LT. 1) GOTO 25
      DO 20 K = 1, L
   20 G = G + Z(I, K) ** 2
   25 H = G + F * F
C
C         IF G IS TOO SMALL FOR ORTHOGONALITY TO BE GUARANTEED,
C         THE TRANSFORMATION IS SKIPPED
C
      IF (G .GT. TOL) GOTO 30
      E(I) = F
      D(I) = ZERO
      GOTO 65
   30 L = L + 1
      G = ZSQRT(H)
      IF (F .GE. ZERO) G = -G
      E(I) = G
      H = H - F * G
      Z(I, I - 1) = F - G
      F = ZERO
      DO 50 J = 1, L
      Z(J, I) = Z(I, J) / H
      G = ZERO
```

```
C
C            FORM ELEMENT OF A * U OF MARTIN
C
      DO 40 K = 1, J
   40 G = G + Z(J, K) * Z(I, K)
      IF (J .GE. L) GOTO 47
      J1 = J + 1
      DO 45 K = J1, L
   45 G = G + Z(K, J) * Z(I, K)
C
C         FORM ELEMENT OF P OF MARTIN
C
   47 E(J) = G / H
      F = F + G * Z(J, I)
   50 CONTINUE
C
C         FORM K
C
      HH = F / (H + H)
C
C         FORM REDUCED A OF MARTIN
C
      DO 60 J = 1, L
      F = Z(I, J)
      G = E(J) - HH * F
      E(J) = G
      DO 60 K = 1, J
      Z(J, K) = Z(J, K) - F * E(K) - G * Z(I, K)
   60 CONTINUE
      D(I) = H
   65 I = I - 1
   70 CONTINUE
      D(1) = ZERO
      E(1) = ZERO
C
C         ACCUMULATION OF TRANSFORMATION MATRICES
C
      DO 110 I = 1, N
      L = I - 1
      IF (D(I) .EQ. ZERO .OR. L .EQ. 0) GOTO 100
      DO 90 J = 1, L
      G = ZERO
      DO 80 K = 1, L
   80 G = G + Z(I, K) * Z(K, J)
      DO 90 K = 1, L
      Z(K, J) = Z(K, J) - G * Z(K, I)
   90 CONTINUE
  100 D(I) = Z(I, I)
      Z(I, I) = ONE
      IF (L .EQ. 0) GOTO 110
      DO 105 J = 1, L
      Z(I, J) = ZERO
      Z(J, I) = ZERO
  105 CONTINUE
  110 CONTINUE
      RETURN
      END
C
      SUBROUTINE LRVT(N, PRECIS, D, E, Z, IFAULT)
C
C         ALGORITHM AS 60.2   APPL. STATIST. (1973) VOL.22, P.260
C
C         FINDS LATENT ROOTS AND VECTORS OF TRIDIAGONAL MATRIX
C
      REAL B, C, D(N), E(N), F, G, H, P, PR, PRECIS, R, S,
     $  Z(N, N), ZERO, ONE, TWO, ZABS, ZSQRT
C
      DATA MITS, ZERO, ONE, TWO /30, 0.0, 1.0, 2.0/
C
```

```
      ZABS(B) = ABS(B)
      ZSQRT(B) = SQRT(B)
C
      IFAULT = 2
      IF (N .LE. 1) RETURN
      IFAULT = 1
      N1 = N - 1
      DO 10 I = 2, N
   10 E(I - 1) = E(I)
      E(N) = ZERO
      B = ZERO
      F = ZERO
      DO 90 L = 1, N
      JJ = 0
      H = PRECIS * (ZABS(D(L)) + ZABS(E(L)))
      IF (B .LT. H) B = H
C
C         LOOK FOR SMALL SUB-DIAGONAL ELEMENT
C
      DO 20 M1 = L, N
      M = M1
      IF (ZABS(E(M)) .LE. B) GOTO 30
   20 CONTINUE
   30 IF (M .EQ. L) GOTO 85
   40 IF (JJ .EQ. MITS) RETURN
      JJ = JJ + 1
C
C         FORM SHIFT
C
      P = (D(L + 1) - D(L)) / (TWO * E(L))
      R = ZSQRT(P * P + ONE)
      PR = P + R
      IF (P .LT. ZERO) PR = P - R
      H = D(L) - E(L) / PR
      DO 50 I = L, N
   50 D(I) = D(I) - H
      F = F + H
C
C         QL TRANSFORMATION
C
      P = D(M)
      C = ONE
      S = ZERO
      M1 = M - 1
      I = M
      DO 80 I1 = L, M1
      J = I
      I = I - 1
      G = C * E(I)
      H = C * P
      IF (ZABS(P) .GE. ZABS(E(I))) GOTO 60
      C = P / E(I)
      R = ZSQRT(C * C + ONE)
      E(J) = S * E(I) * R
      S = ONE / R
      C = C / R
      GOTO 70
   60 C = E(I) / P
      R = ZSQRT(C * C + ONE)
      E(J) = S * P * R
      S = C / R
      C = ONE / R
   70 P = C * D(I) - S * G
      D(J) = H + S * (C * G + S * D(I))
C
C         FORM VECTOR
C
      DO 80 K = 1, N
      H = Z(K, J)
```

```
        Z(K, J) = S * Z(K, I) + C * H
        Z(K, I) = C * Z(K, I) - S * H
   80 CONTINUE
        E(L) = S * P
        D(L) = C * P
        IF (ZABS(E(L)) .GT. B) GOTO 40
   85 D(L) = D(L) + F
   90 CONTINUE
C
C          ORDER LATENT ROOTS AND VECTORS
C
        DO 120 I = 1, N1
        K = I
        P = D(I)
        I1 = I + 1
        DO 100 J = I1, N
        IF (D(J) .GE. P) GOTO 100
        K = J
        P = D(J)
  100 CONTINUE
        IF (K .EQ. I) GOTO 120
        D(K) = D(I)
        D(I) = P
        DO 110 J = 1, N
        P = Z(J, I)
        Z(J, I) = Z(J, K)
        Z(J, K) = P
  110 CONTINUE
  120 CONTINUE
        IFAULT = 0
        RETURN
        END
```

Algorithm AS 63

THE INCOMPLETE BETA INTEGRAL

By K. L. Majumder[†] and G. P. Bhattacharjee

Department of Mathematics,
Indian Institute of Technology,
Kharagpur 721302, India

Keywords: Probability integral; Beta distribution; Series approximation.

LANGUAGE

Fortran 66 and 77.

DESCRIPTION AND PURPOSE

For given values of x ($0 \leqslant x \leqslant 1$), p (>0), q (>0) and the logarithm of the complete beta function, $\ln B(p,q)$, the function subprogram computes the incomplete beta function ratio defined by

$$I_x(p,q) = \frac{1}{B(p,q)} \int_0^x t^{p-1}(1-t)^{q-1}dt.$$

NUMERICAL METHOD

The subprogram uses the method, discussed by Soper (1921). If p is not less than $(p + q)x$ and the integral part of $q + (1 - x)(p + q)$ is a positive integer, say s, reductions are made up to s times 'by parts' using the recurrence relation

$$I_x(p,q) = \frac{\Gamma(p+q)}{\Gamma(p+1)\,\Gamma(q)} \; x^p(1-x)^{q-1} + I_x(p+1, q-1)$$

and then reductions are continued by 'raising p' with the recurrence relation

$$I_x(p+s,\, q-s) = \frac{\Gamma(p+q)}{\Gamma(p+s+1)\,\Gamma(q-s)} \; x^{p+s}(1-x)^{q-s} + I_x(p+s+1, q-s).$$

If s is not a positive integer, reductions are made only by 'raising p'. The process of reduction is terminated when the relative contribution to the integral is not greater than the value of *ACU* defined in the *DATA* statement of the

---

[†]*Present address:* Space Applications Centre, SAC P.O., Ahmedabad – 380 053, India.

subprogram. If p is less than $(p + q)x$, $I_{1-x}(q,p)$ is first calculated by the above procedure and then $I_x(p,q)$ is obtained from the relation

$$I_x(p,q) = 1 - I_{1-x}(q, p).$$

Soper (1921) demonstrated that the expansion of $I_x(p,q)$ by 'parts' and 'raising p' method as described above converges more rapidly than any other series expansions.

STRUCTURE

REAL FUNCTION BETAIN (X, P, Q, BETA, IFAULT)

Formal parameters

X	Real	input:	the value of the upper limit x.
P	Real	input:	the value of the parameter p.
Q	Real	input:	the value of the parameter q.
BETA	Real	input:	the value of $\ln B(p,q)$.
IFAULT	Integer	output:	a fault indicator, equal to: 1 if $p \leqslant 0$ or $q \leqslant 0$; 2 if $x < 0$ or $x > 1$; 0 otherwise.

AUXILIARY ALGORITHM

The algorithm of Pike and Hill (1966), which computes the natural logarithm of the complete gamma function $\Gamma(p)$ for p strictly positive, may be used to calculate $\ln B(p,q)$ from the relation

$$\ln B(p,q) = \ln \Gamma(p) + \ln \Gamma(q) - \ln \Gamma(p + q)$$

ACCURACY

The accuracy of the result should not be less than the value of *ACU* defined in the *DATA* statement of the subprogram; but owing to truncation and rounding errors the last digit may not be correct.

PRECISION

To convert the routine to double precision the following changes should be made:

(1) Change *REAL* to *DOUBLE PRECISION* on both the *FUNCTION* statement and the declaration of variables.
(2) Set constants to double precision values in the *DATA* statement.

(3) Change *ABS, EXP* and *ALOG* to *DABS, DEXP* and *DLOG* respectively in the statement functions.

ACKNOWLEDGEMENT

The authors wish to thank Dr I. D. Hill and the referee for their helpful comments on an earlier version of the algorithm, and the authors of Remark AS R19.

EDITORS' REMARKS

The corrections suggested in Remark AS R19 (Cran *et al.*, 1977) have been applied. The principal effect of these changes is to require ln $B(p,q)$ rather than $B(p,q)$ as input parameter *BETA*. The only other changes made are those facilitating the change to double precision outlined above.

REFERENCES

Cran, G. W., Martin, K. J. and Thomas, G. E. (1977) Remark AS R19 and Algorithm AS 109. *Appl. Statist.*, **26**, 111–114.
Pike, M. C. and Hill, I. D. (1966) Algorithm 291. Logarithm of the gamma function. *Commun. Ass. Comput. Mach.*, **9**, 684. (See also this book, page 243)
Soper, H. E. (1921) The numerical evaluation of the incomplete beta-function. In *Tract for Computers* No. 7. Cambridge: Cambridge University Press.

```
      REAL FUNCTION BETAIN(X, P, Q, BETA, IFAULT)
C
C         ALGORITHM AS 63  APPL. STATIST. (1973) VOL.22, P.409
C
C         COMPUTES INCOMPLETE BETA FUNCTION RATIO FOR ARGUMENTS
C         X BETWEEN ZERO AND ONE, P AND Q POSITIVE.
C         LOG OF COMPLETE BETA FUNCTION, BETA, IS ASSUMED TO
C         BE KNOWN.
C
      LOGICAL INDEX
      REAL ACU, AI, BETA, CX, ONE, P, PP, PSQ, Q, QQ, RX, TEMP,
     $ TERM, X, XX, ZERO, ZABS, ZEXP, ZLOG
C
C         DEFINE ACCURACY AND INITIALISE
C
      DATA ACU, ONE, ZERO /0.1E-7, 1.0, 0.0/
C
      ZABS(X) = ABS(X)
      ZEXP(X) = EXP(X)
      ZLOG(X) = ALOG(X)
C
      BETAIN = X
C
C         TEST FOR ADMISSIBILITY OF ARGUMENTS
C
      IFAULT = 1
      IF (P .LE. ZERO .OR. Q .LE. ZERO) RETURN
      IFAULT = 2
      IF (X .LT. ZERO .OR. X .GT. ONE) RETURN
      IFAULT = 0
      IF (X .EQ. ZERO .OR. X .EQ. ONE) RETURN
```

```
C
C           CHANGE TAIL IF NECESSARY AND DETERMINE S
C
      PSQ = P + Q
      CX = ONE - X
      IF (P .GE. PSQ * X) GOTO 1
      XX = CX
      CX = X
      PP = Q
      QQ = P
      INDEX = .TRUE.
      GOTO 2
    1 XX = X
      PP = P
      QQ = Q
      INDEX = .FALSE.
    2 TERM = ONE
      AI = ONE
      BETAIN = ONE
      NS = QQ + CX * PSQ
C
C           USE REDUCTION FORMULAE OF SOPER.
C
      RX = XX / CX
    3 TEMP = QQ - AI
      IF (NS .EQ. 0) RX = XX
    4 TERM = TERM * TEMP * RX / (PP + AI)
      BETAIN = BETAIN + TERM
      TEMP = ZABS(TERM)
      IF (TEMP .LE. ACU .AND. TEMP .LE. ACU * BETAIN) GOTO 5
      AI = AI + ONE
      NS = NS - 1
      IF (NS .GE. 0) GOTO 3
      TEMP = PSQ
      PSQ = PSQ + ONE
      GOTO 4
C
C           CALCULATE RESULT
C
    5 BETAIN = BETAIN * ZEXP(PP * ZLOG(XX) + (QQ - ONE) *
   $   ZLOG(CX) - BETA) / PP
      IF (INDEX) BETAIN = ONE - BETAIN
      RETURN
      END
```

Algorithm AS 64/AS 109

INVERSE OF THE INCOMPLETE BETA FUNCTION RATIO

By K. L. Majumder[†] and G. P. Bhattacharjee

Department of Mathematics,
Indian Institute of Technology,
Kharagpur 721302, India

Keywords: Probability points; Beta distribution; Newton–Raphson method.

LANGUAGE

Fortran 66 and 77.

DESCRIPTION AND PURPOSE

For given values of $\alpha(0 \leqslant \alpha \leqslant 1)$, $p(>0)$, $q(>0)$ and the logarithm of the complete beta function, $\ln B(p,q)$, the subprogram computes the value of x satisfying the relation

$$I_x(p,q) = \frac{1}{B(p,q)} \int_0^x t^{p-1}(1-t)^{q-1}dt = \alpha, \ 0 \leqslant x \leqslant 1.$$

NUMERICAL METHOD

An approximation x_0 to x is found using the approximation of Carter (1947) if p and $q > 1$, and otherwise from (cf. Scheffé and Tukey, 1944)

$$(1 + x_0)/(1 - x_0) = (4p + 2q - 2)/\chi_\alpha^2,$$

where χ_α^2 is the upper $100\alpha\%$ point of the χ^2 distribution with $2q$ degrees of freedom and is obtained from Wilson and Hilferty's approximation (cf. Wilson and Hilferty, 1931)

$$\chi_\alpha^2 = 2q\left(1 - \frac{1}{9q} + y_\alpha\sqrt{\frac{1}{9q}}\right)^3,$$

y_α being Hastings's approximation (cf. Hastings, 1955) for the upper $100\alpha\%$ point of the standardized normal distribution. If $\chi_\alpha^2 < 0$, then

$$x_0 = 1 - \{(1 - \alpha)qB(p,q)\}^{1/q}.$$

[†]*Present address:* Space Applications Centre, SAC P.O., Ahmedabad – 380 053, India.

Again if $(4p + 2q - 2)/\chi_\alpha^2$ does not exceed 1, x_0 is obtained from

$$x_0 = \{\alpha p B(p,q)\}^{1/p}.$$

The final solution is obtained by the Newton–Raphson method from the relation

$$x_i = x_{i-1} - f(x_{i-1})/f'(x_{i-1}),$$

where

$$f(x) = I_x(p,q) - \alpha.$$

Steps are taken to ensure that x remains in the range $[0, 1]$, and that the value of the adjustment to x, if of the opposite sign to its previous value, is smaller in magnitude. The iteration process is terminated when the square of the absolute difference between two successive iterations is not greater than the accuracy *ACU* defined in the *DATA* statement of the subprogram.

STRUCTURE

REAL FUNCTION XINBTA (P, Q, BETA, ALPHA, IFAULT)

Formal parameters

P	Real	input:	the value of parameter p.
Q	Real	input:	the value of parameter q.
BETA	Real	input:	the value of ln $B(p,q)$.
ALPHA	Real	input:	the value of the lower tail area.
IFAULT	Integer	output:	a fault indicator, equal to: 1 if $p \leqslant 0$ or $q \leqslant 0$; 2 if $\alpha < 0$ or $\alpha > 1$; 3 if during computation *XINBTA* becomes negative or greater than 1; 0 otherwise.

AUXILIARY ALGORITHMS

XINBTA calls the function *BETAIN (X, P, Q, BETA, IFAULT)* for the calculation of the incomplete beta function ratio for which Algorithm AS 63 may be used. The algorithm of Pike and Hill (1966), which computes the natural logarithm of the complete gamma function $\Gamma(p)$ for p strictly positive, may be used to calculate ln $B(p,q)$ from the relation

$$\ln B(p,q) = \ln \Gamma(p) + \ln \Gamma(q) - \ln \Gamma(p + q)$$

ACCURACY

The accuracy of the answer is determined by the value of *ACU* defined in the *DATA* statement of the subprogram.

In general, the exponent of *ACU* should be set at $-2r - 2$, where r is the number of decimal places' accuracy required. Thus the value used below should give accuracy to about six decimal places.

PRECISION

To convert the routine to double precision the following changes should be made:

(1) Change *REAL* to *DOUBLE PRECISION* on both the *FUNCTION* statement and the declaration of variables.
(2) Set constants to double precision values in the *DATA* statement.
(3) Change *EXP, ALOG* and *SQRT* to *DEXP, DLOG* and *DSQRT* respectively in the statement functions.

ACKNOWLEDGEMENT

The authors wish to thank Dr I. D. Hill and the referee for their helpful comments on an earlier version of the algorithm, and the authors of Remark AS R19.

EDITORS' REMARKS

The description presented here is essentially that of the original AS 64, amended in accordance with the suggestions in Remark AS R19 (Cran *et al.*, 1977). The principal change is the requirement for ln $B(p,q)$ rather than $B(p,q)$ as input parameter *BETA*.

The actual Fortran coding is that presented in Cran *et al.*, (1977), which was given the identifier AS 109 in view of the extent of the modification of the original AS 64 code. The only modifications made to AS 109 are those facilitating the change to double precision outlined above.

REFERENCES

Carter, A. H. (1947) Approximation to percentage points of the *z*-distribution. *Biometrika,* **34**, 352–358.

Cran, G. W., Martin, K. J. and Thomas, G. E. (1977) Remark AS R19 and Algorithm AS 109. *Appl. Statist.*, **26**, 111–114.

Hastings, C. J. (1955) *Approximations for Digital Computers.* Princeton: Princeton University Press.

Majumder, K. L. and Bhattacharjee, G. P. (1973) Algorithm AS 63. The incomplete beta integral. *Appl. Statist.*, **22**, 409–411. (See also this book, page 117).

Pike, M. C. and Hill, I. D. (1966) Algorithm 291. Logarithm of the gamma function. *Commun. Ass. Comput. Mach.*, **9**, 684. (See also this book, page 243).

Scheffé, H. and Tukey, J. W. (1944) A formula for small sizes for population tolerance limits. *Ann. Math. Statist.*, **15**, 217.

Wilson, E. B. and Hilferty, M. M. (1931) The distribution of chi-square. *Proc. Nat. Acad. Sci.*, **17**, 684–688.

```
      REAL FUNCTION XINBTA(P, Q, BETA, ALPHA, IFAULT)
C
C         ALGORITHM AS 109  APPL. STATIST. (1977) VOL.26, P.111
C         (REPLACING ALGORITHM AS 64  APPL. STATIST. (1973),
C         VOL.22, P.411)
C
C         COMPUTES INVERSE OF INCOMPLETE BETA FUNCTION
C         RATIO FOR GIVEN POSITIVE VALUES OF THE ARGUMENTS
C         P AND Q, ALPHA BETWEEN ZERO AND ONE.
C         LOG OF COMPLETE BETA FUNCTION, BETA, IS ASSUMED TO BE
C         KNOWN.
C
      LOGICAL INDEX
      REAL A, ALPHA, ADJ, BETA, G, H, P, PP, PREV, Q, QQ, R, S,
     $  SQ, T, TX, W, Y, YPREV, ZERO, HALF, ONE, TWO, THREE,
     $  FOUR, FIVE, SIX, NINE, ACU, LOWER, UPPER, CONST1,
     $  CONST2, CONST3, CONST4, BETAIN, ZEXP, ZLOG, ZSQRT
C
C         DEFINE ACCURACY AND INITIALIZE
C
      DATA ZERO, HALF, ONE, TWO, THREE, FOUR, FIVE, SIX, NINE
     $     /0.0,  0.5, 1.0, 2.0,   3.0,  4.0,  5.0, 6.0,  9.0/
      DATA   ACU,  LOWER,  UPPER,   CONST1,  CONST2,  CONST3,  CONST4
     $  /1.0E-14, 0.0001, 0.9999, 2.30753, 0.27061, 0.99229, 0.04481/
C
      ZEXP(A) = EXP(A)
      ZLOG(A) = ALOG(A)
      ZSQRT(A) = SQRT(A)
C
      XINBTA = ALPHA
C
C         TEST FOR ADMISSIBILITY OF PARAMETERS
C
      IFAULT = 1
      IF (P .LE. ZERO .OR. Q .LE. ZERO) RETURN
      IFAULT = 2
      IF (ALPHA .LT. ZERO .OR. ALPHA .GT. ONE) RETURN
      IFAULT = 0
      IF (ALPHA .EQ. ZERO .OR. ALPHA .EQ. ONE) RETURN
C
C         CHANGE TAIL IF NECESSARY
C
      IF (ALPHA .LE. HALF) GOTO 1
      A = ONE - ALPHA
      PP = Q
      QQ = P
      INDEX = .TRUE.
      GOTO 2
    1 A = ALPHA
      PP = P
      QQ = Q
      INDEX = .FALSE.
C
C         CALCULATE THE INITIAL APPROXIMATION
C
    2 R = ZSQRT(-ZLOG(A * A))
      Y = R - (CONST1 + CONST2 * R) / (ONE + (CONST3 + CONST4 *
     $  R) * R)
      IF (PP .GT. ONE .AND. QQ .GT. ONE) GOTO 5
      R = QQ + QQ
      T = ONE / (NINE * QQ)
      T = R * (ONE - T + Y * ZSQRT(T)) ** 3
      IF (T .LE. ZERO) GOTO 3
      T = (FOUR * PP + R - TWO) / T
      IF (T .LE. ONE) GOTO 4
      XINBTA = ONE - TWO / (T + ONE)
      GOTO 6
    3 XINBTA = ONE - ZEXP((ZLOG((ONE - A) * QQ) + BETA) / QQ)
      GOTO 6
```

```
    4 XINBTA = ZEXP((ZLOG(A * PP) + BETA) / PP)
      GOTO 6
    5 R = (Y * Y - THREE) / SIX
      S = ONE / (PP + PP - ONE)
      T = ONE / (QQ + QQ - ONE)
      H = TWO / (S + T)
      W = Y * ZSQRT(H + R) / H - (T - S) * (R + FIVE / SIX -
    $ TWO / (THREE * H))
      XINBTA = PP / (PP + QQ * ZEXP(W + W))
C
C         SOLVE FOR X BY A MODIFIED NEWTON-RAPHSON METHOD,
C         USING THE FUNCTION BETAIN
C
    6 R = ONE - PP
      T = ONE - QQ
      YPREV = ZERO
      SQ = ONE
      PREV = ONE
      IF (XINBTA .LT. LOWER) XINBTA = LOWER
      IF (XINBTA .GT. UPPER) XINBTA = UPPER
    7 Y = BETAIN(XINBTA, PP, QQ, BETA, IFAULT)
      IF (IFAULT .EQ. 0) GOTO 8
      IFAULT = 3
      RETURN
    8 Y = (Y - A) * ZEXP(BETA + R * ZLOG(XINBTA) + T *
    $ ZLOG(ONE - XINBTA))
      IF (Y * YPREV .LE. ZERO) PREV = SQ
      G = ONE
    9 ADJ = G * Y
      SQ = ADJ * ADJ
      IF (SQ .GE. PREV) GOTO 10
      TX = XINBTA - ADJ
      IF (TX .GE. ZERO .AND. TX .LE. ONE) GOTO 11
   10 G = G / THREE
      GOTO 9
   11 IF (PREV .LE. ACU) GOTO 12
      IF (Y * Y .LE. ACU) GOTO 12
      IF (TX .EQ. ZERO .OR. TX .EQ. ONE) GOTO 10
      IF (TX .EQ. XINBTA) GOTO 12
      XINBTA = TX
      YPREV = Y
      GOTO 7
   12 IF (INDEX) XINBTA = ONE - XINBTA
      RETURN
      END
```

Algorithm AS 66

THE NORMAL INTEGRAL
By I. D. Hill

MRC Clinical Research Centre, Watford Road, Harrow, Middlesex, HA1 3UJ, UK

Keywords: Normal curve; Tail area.

LANGUAGE

Fortran 66 and 77.

DESCRIPTION AND PURPOSE

Calculates the upper, or lower, tail area of the standardized normal curve corresponding to any given argument.

NUMERICAL METHOD

The method is that of Adams (1969), but incorporated in a different surrounding structure.

STRUCTURE

REAL FUNCTION ALNORM (X, UPPER)

Formal parameters

X	Real	input:	the argument value.
UPPER	Logical	input:	if .*TRUE.* the area found is from X to infinity, if .*FALSE.* from minus infinity to X.

Data constants

The constant *LTONE* should be set to the value at which the lower tail area becomes 1·0 to the accuracy of the machine. *LTONE* = $(n + 9)/3$ gives the required value accurately enough, for a machine that produces n decimal digits in its real numbers.

The constant *UTZERO* should be set to the value at which the upper tail area becomes 0·0 to the accuracy of the machine. This may be taken as

$$\sqrt{(-2(\ln x + 1))} - 0·3$$

where x is the smallest allowable real number.

RELATED ALGORITHMS

Algorithm AS 2 (Cooper, 1968) exists for the same purpose, but, as Hitchin (1973) has shown, is not always an easy algorithm to use in that both its input and output parameters have to be in array form.

It is believed that the scalar arguments, and functional form, of the current algorithm are much simpler to use, and it is considerably faster.

ACCURACY

About nine significant figures (decimal) are correct on a machine that works to such accuracy.

PRECISION

For a double precision version:

(1) Change *REAL FUNCTION* to *DOUBLE PRECISION FUNCTION.*
(2) Change *REAL* to *DOUBLE PRECISION.*
(3) Change the constants in the *DATA* statements to double precision values and increase the value of *LTONE* to correspond to the new precision.
(4) Change *EXP* to *DEXP* in the statement function.

ADDITIONAL COMMENTS

For full accuracy it is important to use the *UPPER* parameter for the required tail area, and not to resort to

$$1 \cdot 0 - ALNORM(X, .TRUE.)$$

instead of

$$ALNORM(X, .FALSE.)$$

since the former can lead to a severe loss of significant figures.

Expressions may, of course, be used for either parameter. To many users it seems to come naturally to use an expression for a numerical variable, but not for a logical one, yet the latter can sometimes be useful. For instance

$$ALNORM(X, X .GT. 0.0)$$

can be used to find the smaller of the two tail areas, and twice this value is needed for a two-tail test.

ACKNOWLEDGEMENT

I am grateful to Dr J. A. Nelder for suggesting a number of improvements in presentation.

EDITORS' REMARKS

The only changes made to this routine are those facilitating the change to double precision outlined above.

REFERENCES

Adams, A. G. (1969) Algorithm 39. Areas under the normal curve. *Computer J.*, **12**, 197–198.

Cooper, B. E. (1968) Algorithm AS 2. The normal integral. *Appl. Statist.*, **17**, 186–187.

Hill, I. D. (1969) Remark AS R2. A remark on algorithm AS 2. *Appl. Statist.*, **18**, 299–300.

Hitchin, D. (1973) Remark AS R8. A remark on algorithms AS 4 and AS 5. *Appl. Statist.*, **22**, 428.

```
      REAL FUNCTION ALNORM(X, UPPER)
C
C        ALGORITHM AS 66  APPL. STATIST. (1973) VOL.22, P.424
C
C        EVALUATES THE TAIL AREA OF THE STANDARDIZED NORMAL CURVE
C        FROM X TO INFINITY IF UPPER IS .TRUE. OR
C        FROM MINUS INFINITY TO X IF UPPER IS .FALSE.
C
      REAL LTONE, UTZERO, ZERO, HALF, ONE, CON, A1, A2, A3,
     $   A4, A5, A6, A7, B1, B2, B3, B4, B5, B6, B7, B8, B9,
     $   B10, B11, B12, X, Y, Z, ZEXP
      LOGICAL UPPER, UP
C
C        LTONE AND UTZERO MUST BE SET TO SUIT THE PARTICULAR COMPUTER
C        (SEE INTRODUCTORY TEXT)
C
      DATA LTONE, UTZERO /7.0, 18.66/
      DATA ZERO, HALF, ONE, CON /0.0, 0.5, 1.0, 1.28/
      DATA          A1,              A2,              A3,
     $              A4,              A5,              A6,
     $              A7
     $   /0.398942280444, 0.399903438504, 5.75885480458,
     $      29.8213557808, 2.62433121679, 48.6959930692,
     $      5.92885724438/
      DATA          B1,              B2,              B3,
     $              B4,              B5,              B6,
     $              B7,              B8,              B9,
     $              B10,             B11,             B12
     $   /0.398942280385,    3.8052E-8,  1.00000615302,
     $      3.98064794E-4, 1.98615381364, 0.151679116635,
     $      5.29330324926, 4.8385912808,  15.1508972451,
     $      0.742380924027, 30.789933034, 3.990194170011/
C
      ZEXP(Z) = EXP(Z)
C
      UP = UPPER
      Z = X
      IF (Z .GE. ZERO) GOTO 10
      UP = .NOT. UP
      Z = -Z
   10 IF (Z .LE. LTONE .OR. UP .AND. Z .LE. UTZERO) GOTO 20
      ALNORM = ZERO
      GOTO 40
```

```
   20 Y = HALF * Z * Z
      IF (Z .GT. CON) GOTO 30
C
      ALNORM = HALF - Z * (A1 - A2 * Y / (Y + A3 - A4 / (Y + A5 +
     $ A6 / (Y + A7))))
      GOTO 40
C
   30 ALNORM = B1 * ZEXP(-Y) / (Z - B2 + B3 / (Z + B4 + B5 / (Z -
     $ B6 + B7 / (Z + B8 - B9 / (Z + B10 + B11 / (Z + B12)))))
C
   40 IF (.NOT. UP) ALNORM = ONE - ALNORM
      RETURN
      END
```

Algorithm AS 75

BASIC PROCEDURES FOR LARGE, SPARSE OR WEIGHTED LINEAR LEAST SQUARES PROBLEMS

By W. Morven Gentleman

University of Waterloo

Present address: Division of Electrical Engineering, National Research Council of Canada, Ottawa, K1A 0R8, Canada.

Keywords: Linear least squares; Givens transformations; Plane rotations.

LANGUAGES

Algol 60, and Fortran 66 and 77.

DESCRIPTION AND PURPOSE

Forming and solving the normal equations numerically is often an unsatisfactory way to solve linear least squares problems, partly because of the loss of accuracy which results when the crossproduct matrix is formed (Longley, 1967; Wampler, 1970), and partly because adding or deleting observations is often required, and it is desirable to take advantage of arithmetic already done. Orthogonal decomposition methods based on the modified Gram Schmidt algorithm (*not* the classical one) or based on Householder transformations overcome the accuracy problem, but they are not well suited to updating the solution to include or exclude observations. Moreover, they are typically programmed to require the whole $n \times p$ design matrix to be in high-speed store, thus restricting the size of problem for which they can be used, and they cannot conveniently be implemented so as to exploit sparseness in the design matrix.

Orthogonal decomposition methods based on Givens transformations (plane rotations) are as accurate as any other orthogonal decomposition method, but have two major advantages: zeros in the design matrix are readily exploited in obvious ways for substantial gains, and the design matrix can naturally be processed by rows (observations). Indeed, one way to view methods based on Givens transformations is as numerically stable ways to update the Cholesky (matrix square root) decomposition of the crossproduct matrix to include one more observation, from which it is clear that only the $p \times p$ triangular Cholesky factor and the new row need be in high-speed store. Givens transformations have been used in the least squares problem by Fowlkes (1969) and Chambers (1971), and in the linear programming problem by Saunders (1972).

An improved version of this approach is developed in Gentleman (1973). By inserting a diagonal scaling matrix between the factors of the Cholesky decomposition, it is possible to enjoy all the advantages of the usual version and (1) to avoid all square roots (Givens transformations normally require one square root for each element of the design matrix), (2) to halve the amount of arithmetic required (Givens transformations normally require twice as much

arithmetic in the dense case as Gram Schmidt or Householder) and (3) to introduce each new row with arbitrary positive or negative weight (thus solving
weighted least squares problems or deleting observations from a regression).

Four of the procedures given here form the basic set of procedures needed
to use this method for linear regression analysis: they include new observations
in the regression, find confounded contrasts causing rank deficiencies, obtain
the sum of squares decomposition and find the regression coefficients. The fifth
procedure illustrates a typical, if rather simplified, use of the other four: it
provides an analysis of factorial designs for which no other program is available,
or which have become unbalanced because of either weighting, missing or
extra observations, or perhaps the presence of covariates.

STRUCTURE

procedure *include (p, weight, X row, y element, D, R bar, theta bar, SS error)*
which updates *D, R bar, theta bar,* and *SS error* to include, with specified
weight, the effect of a new observation *X row* and *y element.* For an initial
decomposition *D, R bar, theta bar,* and *SS error* should be set to zero before
including the first row. Alternatively, these quantities can be initialized by a
preliminary decomposition of some of the observations of the regression, using,
for example the square root-free version of modified Gram Schmidt (Clayton,
1971) or the square root-free version of the Cholesky decomposition applied to
a crossproduct matrix (Martin *et al.,* 1965; Healy, 1968).

procedure *confound (p, j, R bar, contrast)*
which, given $\bar{R}$ and some integer *j,* finds the contrast which could not be estimated if D_{jj} were zero, that is, finds the linear combination of the first *j* columns
of *X* which would vanish (Fowlkes, 1969). Most cases where *X* is not of full
rank (that is, where the independent variables are confounded) can readily be
detected by some D_{jj} becoming small or vanishing. The standard method of
resolving the resulting indeterminacy is to find the confounded contrast (as
produced by this procedure), and then either to force one of the confounded
variables (those with non-zero coefficients in the contrast) to have regression
coefficient zero, or to orthogonalize the regression coefficients of a subset of
confounded variables to the others. The latter is achieved by requiring the
vanishing of a linear combination of regression coefficients equal to the confounded contrast for the components in the subset, and zero for the other
components. Constraints like either of the above, which merely resolve indeterminacy, can readily be imposed by including them as extra observations *X row*
and *y.*[†]

[†]If a rank deficient matrix is decomposed by a QR decomposition, the typical behaviour is
that a very small element appears on the diagonal, and all the elements of *R* to the right or
below this vary unpredictably depending on the specific rounding arithmetic of the machine
used. This is not, however, evidence of numerical instability, for the elements produced
are related in such a way that when the rank degeneracy represented by the small diagonal
element is resolved, e.g. in the manner suggested above, the elements of *R* become well
determined and accurately computed, up to the next very small element on the diagonal.

procedure *SS decomp (p, D, theta bar, SS)*
which, given D and $\bar{\theta}$, computes the sum of squares decompositions. This, and not the regression coefficients, is what is needed for standard hypothesis testing.

procedure *regress (p, R bar, theta bar, beta)*
which, given $\bar{R}$ and $\bar{\theta}$, computes the regression coefficients *beta*.

procedure *analyze (input, output)*
which uses procedures *ininteger* and *inreal* to read data from channel *input,* and uses procedures *outstring, outinteger, outreal, outarray, newline* and *newpage* to produce the results of its analysis on channel *output.* (These input—output procedures, except for the obvious procedures *newline* and *newpage,* conform to the IFIP standard as extended by the *Comm. ACM* Algorithms policy). The input medium is assumed to contain first a value for the integer p, then a sequence of records containing the rows of X and y in compact form, together with corresponding weights, and finally the number zero as a flag to indicate the end of the data. The compact form of each row consists first of the weight, then the y element, then pairs consisting of the column index of non-zero elements in this row of X followed by the value of the element. A column index of zero indicates that there are no further non-zero elements in that row. The tests used to detect rank deficiencies and to identify the confounded variables (10^{-8} and 10^{-4} respectively) are obviously oversimplified; and more sensible tests could of course be used (Healy, 1968; Clayton, 1971). Similarly, the resolving constraints defined are obviously oversimplified, and are only appropriate in the absence of interactions. Nevertheless, even with these gross simplifications, the analyses of most standard designs without interactions are correctly computed by this procedure on the IBM 360, which works to an accuracy of six digits.

Formal parameters

		include	confound	SS decomp	regress	analyze
p	Integer	value	value	value	value	read
weight	Real	value				read
X row	Real array [1: p]	input				read
y element	Real	value				read
D	Real array [1: p]	input/ output		input		
R bar	Real array [1: $p \times (p-1) \div 2$]	input/ output	input		input	
theta bar	Real array [1: p]	input/ output		input	input	
SS error	Real	input/ output				
j	Integer		value			
contrast	Real array [1: p]		output			
SS	Real array [1: p]			output		
beta	Real array [1: p]				output	
input	Integer					value
output	Integer					value

p — the number of independent variables

weight — the weight of this observation in the regression

X *row* — the independent variables for this observation

y element — the dependent variable for this observation

D — the diagonal scaling matrix

R *bar* — the superdiagonal elements of $\bar{R}$, stored sequentially by rows. $\bar{R}$ is the unit upper triangular matrix such that $D^{1/2}\,\bar{R}$ is the Cholesky triangular factor

theta bar — $\bar{\theta}$, where $D^{1/2}\,\bar{\theta}$ is the vector of orthogonal coefficients

SS error — the sum of squares error

j — see description of *confound* above

contrast — the coefficients of the confounded contrast among the independent variables if the system is rank deficient

SS — the sum of squares decomposition, i.e. the squares of the orthogonal coefficients

beta — the regression coefficients

input — the input channel number

output — the output channel number

RESTRICTIONS

Removing observations from a regression is inherently a numerically unstable process, and if procedure *include* is used with negative weights to do this, then monitoring code should be inserted to detect the instability and restart the decomposition if necessary.

TIME

For a fully dense design matrix, including one more observation by this method takes $\sim p^2$ multiplications and additions. Using Princeton modified Algol on the IBM 360/75, this type of behaviour is indeed observed, with the empirical cost of $46p^2 + 716p - 2591$ microseconds per observation. (This includes indexing, procedure call overhead, etc.) On sparse design matrices, savings of up to 70 per cent have been measured.

ACCURACY

The error analysis of this method (Gentleman, 1973) shows that if non-negative weights are used for all observations the computed solution is the exact solution of a linear least squares problem with a design matrix and observation vector very close to the given ones. In practice the method is often the most accurate single precision method. If this accuracy is not adequate, an iterative improvement can be used (Björck, 1967).

EDITORS' REMARKS

This algorithm has been modified to incorporate the Correction which appeared in 1982, Issue 3. The abbreviation **array** has been changed to **real array** once in each of the five procedures. This does no harm, and may be helpful to anyone wishing to translate to another language. No other changes have been made. Farebrother (1976) proposed a modified version of *include, include2*, which also calculated recursive residuals. That extension to the basic AS 75 is not included, but readers' attention is drawn to its existence, and also to the fact that a Fortran version is included in this book as AS 154.3.

    Fortran versions of these routines are added for compatibility with other algorithms in this book.

    The Algol representation used regards spaces in strings as significant.

REFERENCES

Björck, Å. (1967) Iterative refinement of linear least squares solutions. I., *BIT*, **7**, 257–278.

Clayton, D. G. (1971) Algorithm AS 46. Gram-Schmidt orthogonalization. *Appl. Statist.,* **20**, 335–338.

Chambers, J. M. (1971) Regression updating. *J. Amer. Statist. Ass.,* **66**, 744–748.

Farebrother, R. W. (1976) Remark AS R17. *Appl. Statist.,* **25**, 323–324.

Fowlkes, E. B. (1969) Some operators for ANOVA calculations. *Technometrics,* **11**, 511–526.

Gentleman, W. M. (1973) Least squares computations by Givens transformations without square roots. *J.I.M.A.,* **12**, 329–336.

Healy, M. J. R. (1968) Algorithm AS 6. Triangular decomposition of a symmetric matrix. *Appl. Statist.,* **17**, 195–197. (See also this book, page 43).

Longley, J. W. (1967) An appraisal of least squares programs for the electronic computer from the point of view of the user. *J. Amer. Statist. Ass.,* **62**, 819–841.

Martin, R. S., Peters, G. and Wilkinson, J. H. (1965) Symmetric decomposition of positive definite matrix. *Num. Math.,* **7**, 362–383.

Saunders, M. A. (1972) *Large scale linear programming using the Cholesky factorization.* Technical Report No. CS–72–252, Computer Science Department, Stanford University.

Wampler, R. H. (1970) On the accuracy of least squares computer programs. *J. Amer. Statist. Ass.,* **65**, 549–565.

procedure *include*
(*p, weight, X row, y element, D, R bar, theta bar, SS error*);

comment Algorithm AS 75.1 Appl. Statist. (1974) Vol.23, p.448;

value *p, weight, y element;* **integer** *p;*
real *weight, y element, SS error;* **real array** *X row, D, R bar, theta bar;*

comment invoking this procedure updates *D, R bar, theta bar* and
SS error by the inclusion of *X row, y element* with specified weight;

```
begin real c bar, s bar, xi, xk, di, d prime i; integer i, k, next r;
for i := 1 step 1 until p do
    begin

    comment skip unnecessary transformations. Test on exact zeros
    must be used or stability can be destroyed;

    if weight = 0.0 then goto done;
    if X row[i] ≠ 0.0 then
        begin
        xi := X row[i]; di := D[i];
        d prime i := di + weight × xi ↑ 2; c bar := di / d prime i;
        s bar := weight × xi / d prime i; weight := c bar × weight;
        D[i] := d prime i; next r := (i − 1) × (2 × p − i) ÷ 2 + 1;
        for k := i + 1 step 1 until p do
            begin
            xk := X row[k]; X row[k] := xk − xi × R bar[next r];
            R bar[next r] := c bar × R bar[next r] + s bar × xk;
            next r := next r + 1
            end k;
        xk := y element; y element := xk − xi × theta bar[i];
        theta bar[i] := c bar × theta bar[i] + s bar × xk
        end
    end i;
    SS error := SS error + weight × y element ↑ 2;
done:
    end include;
```

procedure confound(p, j, R bar, contrast);

comment Algorithm AS 75.2 Appl. Statist. (1974) Vol.23, p.448;

value p, j; integer p, j; real array R bar, contrast;

comment invoking this procedure obtains the contrast which could not
be estimated if $D[j]$ were assumed to be zero, that is, obtains the
linear combination of the first j columns which would be zero. This is
obtained by setting the first $j − 1$ elements of contrast to the solution
of the triangular system formed by the first $j − 1$ rows and columns of
R bar with the first $j − 1$ elements of the jth column as right hand side,
setting the jth element of contrast to −1, and setting the remaining
elements of contrast to zero;

```
begin integer i, k, next r;
for i := j + 1 step 1 until p do contrast[i] := 0.0;
contrast[j] := − 1.0;
for i := j − 1 step −1 until 1 do
    begin
    next r := (i − 1) × (2 × p − i) ÷ 2 + 1;
```

```
        contrast[i] := R bar[next r + j − i − 1];
        for k := i + 1 step 1 until j − 1 do
            begin
            contrast[i] := contrast[i] − R bar[next r] × contrast[k];
            next r := next r + 1
            end k
        end i
    end confound;
```

procedure *SS decomp(p, D, theta bar, SS)*;

comment Algorithm AS 75.3 Appl. Statist. (1974) Vol.23, p.448;

value *p*; **integer** *p*; **real array** *D, theta bar, SS*;

comment invoking this procedure computes the *p* components of the sum
of squares decomposition from *D* and *theta bar*;

```
    begin integer i;
    for i := 1 step 1 until p do SS[i] := D[i] × theta bar[i] ↑ 2
    end SS decomp;
```

procedure *regress* (*p, R bar, theta bar, beta*);

comment Algorithm AS 75.4 Appl. Statist. (1974) Vol.23, p.448;

value *p*; **integer** *p*; **real array** *R bar, theta bar, beta*;

comment invoking this procedure obtains *beta* by backsubstitution in
the triangular system *R bar* and *theta bar*;

```
    begin integer i, k, next r;
    for i := p step −1 until 1 do
        begin
        beta[i] := theta bar[i]; next r := (i − 1 ) × (2 × p − i) ÷ 2 + 1;
        for k := i + 1 step 1 until p do
            begin
            beta[i] := beta[i] − R bar[next r] × beta[k];
            next r := next r + 1
            end k
        end i
    end regress;
```

procedure *analyze(input, output)*;

comment Algorithm AS 75.5 Appl. Statist. (1974) Vol.23, p.448;

value *input, output*; **integer** *input, output*;

comment this procedure reads data from channel *input* and produces least squares variance component analyses on channel *output*. The input medium is assumed to contain first a value for the integer p, the number of independent variates, and then a sequence of records containing the rows of X and y in compact form, together with weights. The number zero (read as a weight) will indicate the end of each data set and, upon reading this, an analysis will be produced. Successive analyses will be done until p, the number of independent variates, is read as zero;

```
        begin integer p;
loop:
    ininteger(input, p);
    if p > 0 then
            begin integer n, j, k; real w, y, error; Boolean first;
            real array X, D, theta[1 : p], R[1 : p × (p − 1) ÷ 2];
            n := 0; error := 0.0;
            for k := 1 step 1 until p do D[k] := theta[k] := 0.0;
            for k := p × (p − 1) ÷ 2 step −1 until 1 do R[k] := 0.0;
get: inreal(input, w);
        if w ≠ 0.0 then
            begin
            n := n + sign(w); inreal(input, y);
            for k := 1 step 1 until p do X[k] := 0.0;
next: ininteger(input, k);
            if k > 0 ∧ k ⩽ p then
                begin
                inreal(input, X[k]); goto next
                end;
            include(p, w, X, y, D, R, theta, error); goto get
            end;
        newpage(output); outinteger(output, n);
        outstring(output, 'observations read'); newline(output);
        outstring(output, 'diagonal matrix is'); newline(output);
        outarray(output, D); newline(output);
        first := true;
        for j := 1 step 1 until p do
        if abs(D[j]) < 10−8 then
            begin
```

comment confounding discovered;

```
            if first then
                begin
                first := false; outstring(output, 'confounded contrasts');
                newline(output)
                end;
            confound(p, j, R, X); outarray(output, X);
            newline(output);
```

  comment choose resolving constraint;

  for $k := 1$ **step** 1 **until** $j - 1$ **do**
  if $abs(X[k]) > {}_{10}-4$ **then**
    begin
    $X[k] := 0.0$; **goto** *out*
    end;
 out: *include*(*p*, 1.0, *X*, 0.0, *D*, *R*, *theta*, *error*)
  end *j*;
 outstring(*output*, 'sum of squares decomposition');
 newline(*output*); *SS decomp*(*p*, *D*, *theta*, *X*);
 outarray(*output*, *X*); *newline*(*output*);
 outstring(*output*, 'sum of squares error');
 outreal(*output*, *error*); *newline*(*output*);
 outstring(*output*, 'regression coefficients'); *newline*(*output*);
 regress(*p*, *R*, *theta*, *X*); *outarray*(*output*, *X*);
 goto *loop*
 end
end *analyze*

Fortran version of Algorithm AS 75

STRUCTURE

The Fortran 66 and 77 version follows the original Algol 60 as closely as possible. The various subroutines still function in the same way as their Algol equivalents, but routine names and variable names have had to be changed and truncated, some additional parameters introduced, together with checks on the validity of these extra parameters. For convenience in adjusting the layout of the input and output performed by AS 75.5, each *READ* and *WRITE* statement is allocated its own *FORMAT* statement.

SUBROUTINE INCLUD (NP, NRBAR, WEIGHT, XROW, YELEM, D, RBAR, THETAB, SSERR, IFAULT)
 corresponds to **procedure** *include* (AS 75.1)
SUBROUTINE CONFND (NP, NRBAR, J, RBAR, CONTRA, IFAULT)
 corresponds to **procedure** *confound* (AS 75.2)
SUBROUTINE SSDCMP (NP, D, THETAB, SS, IFAULT)
 corresponds to **procedure** *SS decomp* (AS 75.3)
SUBROUTINE REGRSS (NP, NRBAR, RBAR, THETAB, BETA, IFAULT)
 corresponds to **procedure** *regress* (AS 75.4)
SUBROUTINE ANALYZ (MAXP, MAXR, NP, NRBAR, INPUT, OUTPUT, X, DD, THETA, R, IFAULT)
 corresponds to **procedure** *analyze* (AS 75.5)

Formal parameters

		Algol 60 equivalent	INCLUD	CONFND	SSDCMP	REGRSS	ANALYZ
NP	Integer	p	input	input	input	input	output
NRBAR	Integer	$p \times (p-1)/2$	input	input		input	output
WEIGHT	Real	weight	input				
XROW	Real array (NP)	X row	input/ output				
YELEM	Real	y element	input				
D	Real array (NP)	D	input/ output		input		
RBAR	Real array (NRBAR)	R bar	input/ output	input		input	
THETAB	Real array (NP)	theta bar	input/ output		input	input	
SSERR	Real	SS error	input/ output				
J	Integer	j		input			
CONTRA	Real array (NP)	contrast		output			
SS	Real array (NP)	SS			output		
BETA	Real array (NP)	beta				output	
INPUT	Integer	input					input
OUTPUT	Integer	output					input

Additional parameters

MAXP	Integer		input:	the dimension of X, DD and $THETA$.
MAXR	Integer		input:	the dimension of R.
X	Real array (MAXP)		workspace:	
DD	Real array (MAXP)		workspace:	
THETA	Real array (MAXP)		workspace:	
R	Real array (MAXR)		workspace:	
IFAULT	Integer		output:	a fault indicator, equal to 1 if $NP < 1$ or $NRBAR \neq NP \times (NP-1)/2$ (in INCLUD, CONFND, SSDCMP or REGRSS); 2 if $NP < 0$ or $NP > MAXP$ or $MAXR < NP \times (NP-1)/2$ (in ANALYZ); 3 if an attempt is made to read a value into $X(K)$ where $K < 0$ or $K > NP$ (in ANALYZ); 0 otherwise (all routines).

PRECISION

The routines can be converted into double precision by changing *REAL* to *DOUBLE PRECISION*, replacing the real constants in the *DATA* statements by double precision values throughout, and changing *ABS* to *DABS* in *ANALYZ*.

```
      SUBROUTINE INCLUD(NP, NRBAR, WEIGHT, XROW, YELEM, D, RBAR,
     $ THETAB, SSERR, IFAULT)
C
C        ALGORITHM AS 75.1  APPL. STATIST. (1974) VOL.23, P.448
C
C        CALLING THIS SUBROUTINE UPDATES D, RBAR, THETAB AND SSERR
C        BY THE INCLUSION OF XROW, YELEM WITH SPECIFIED WEIGHT.
C
      REAL XROW(NP), D(NP), RBAR(NRBAR), THETAB(NP), WEIGHT, YELEM,
     $ SSERR, CBAR, DI, DPI, SBAR, W, XI, XK, Y, ZERO
C
      DATA ZERO /0.0/
C
C        CHECK INPUT PARAMETERS
C
      IFAULT = 1
      IF (NP .LT. 1 .OR. NRBAR .NE. NP * (NP - 1) / 2) RETURN
      IFAULT = 0
C
      W = WEIGHT
      Y = YELEM
      DO 100 I = 1, NP
C
C        SKIP UNNECESSARY TRANSFORMATIONS.
C        TEST ON EXACT ZEROS MUST BE USED OR STABILITY CAN BE
C        DESTROYED.
C
      IF (W .EQ. ZERO) RETURN
      IF (XROW(I) .EQ. ZERO) GOTO 100
      XI = XROW(I)
      DI = D(I)
      DPI = DI + W * XI * XI
      CBAR = DI / DPI
      SBAR = W * XI / DPI
      W = CBAR * W
      D(I) = DPI
      IF (I .EQ. NP) GOTO 60
      NEXTR = (I - 1) * (NP + NP - I) / 2 + 1
      IP = I + 1
      DO 50 K = IP, NP
      XK = XROW(K)
      XROW(K) = XK - XI * RBAR(NEXTR)
      RBAR(NEXTR) = CBAR * RBAR(NEXTR) + SBAR * XK
      NEXTR = NEXTR + 1
   50 CONTINUE
   60 XK = Y
      Y = XK - XI * THETAB(I)
      THETAB(I) = CBAR * THETAB(I) + SBAR * XK
  100 CONTINUE
      SSERR = SSERR + W * Y * Y
      RETURN
      END
C
      SUBROUTINE CONFND(NP, NRBAR, J, RBAR, CONTRA, IFAULT)
C
C        ALGORITHM AS 75.2  APPL. STATIST. (1974) VOL.23, P.448
C
```

```
C           CALLING THIS SUBROUTINE OBTAINS THE CONTRAST WHICH COULD
C           NOT BE ESTIMATED IF D(J) WERE ASSUMED TO BE ZERO, THAT IS,
C           OBTAINS THE LINEAR COMBINATION OF THE FIRST J COLUMNS WHICH
C           WOULD BE ZERO.  THIS IS OBTAINED BY SETTING THE FIRST J-1
C           ELEMENTS OF CONTRA TO THE SOLUTION OF THE TRIANGULAR
C           SYSTEM FORMED BY THE FIRST J-1 ROWS AND COLUMNS OF RBAR
C           WITH THE FIRST J-1 ELEMENTS OF THE JTH COLUMN AS RIGHT HAND
C           SIDE, SETTING THE JTH ELEMENT OF CONTRA TO -1, AND SETTING
C           THE REMAINING ELEMENTS OF CONTRA TO ZERO.
C
      REAL RBAR(NRBAR), CONTRA(NP), ZERO, ONE
C
      DATA ZERO, ONE /0.0, 1.0/
C
C           CHECK INPUT PARAMETERS
C
      IFAULT = 1
      IF (NP .LT. 1 .OR. NRBAR .NE. NP * (NP - 1) / 2) RETURN
      IFAULT = 0
C
      JM = J - 1
      IF (J .EQ. NP) GOTO 20
      JP = J + 1
      DO 10 I = JP, NP
   10 CONTRA(I) = ZERO
   20 CONTRA(J) = -ONE
      IF (J .EQ. 1) RETURN
      DO 100 IJ = 1, JM
      I = J - IJ
      NEXTR = (I - 1) * (NP + NP - I) / 2 + 1
      K = NEXTR + J - I - 1
      CONTRA(I) = RBAR(K)
      IF (I .EQ. JM) GOTO 100
      IP = I + 1
      DO 50 K = IP, JM
      CONTRA(I) = CONTRA(I) - RBAR(NEXTR) * CONTRA(K)
      NEXTR = NEXTR + 1
   50 CONTINUE
  100 CONTINUE
      RETURN
      END
C
      SUBROUTINE SSDCMP(NP, D, THETAB, SS, IFAULT)
C
C         ALGORITHM AS 75.3  APPL. STATIST. (1974) VOL.23, P.448
C
C         CALLING THIS SUBROUTINE COMPUTES THE NP COMPONENTS OF
C         THE SUM OF SQUARES DECOMPOSITION FROM D AND THETAB.
C
      REAL D(NP), THETAB(NP), SS(NP)
C
C         CHECK INPUT PARAMETERS
C
      IFAULT = 1
      IF (NP .LT. 1) RETURN
      IFAULT = 0
C
      DO 10 I = 1, NP
   10 SS(I) = D(I) * THETAB(I) ** 2
      RETURN
      END
C
      SUBROUTINE REGRSS(NP, NRBAR, RBAR, THETAB, BETA, IFAULT)
C
C         ALGORITHM AS 75.4  APPL. STATIST. (1974) VOL.23, P.448
C
C         CALLING THIS SUBROUTINE OBTAINS BETA BY BACK-SUBSTITUTION
C         IN THE TRIANGULAR SYSTEM RBAR AND THETAB.
C
```

```
      REAL RBAR(NRBAR), THETAB(NP), BETA(NP)
C
C         CHECK INPUT PARAMETERS
C
      IFAULT = 1
      IF (NP .LT. 1 .OR. NRBAR .NE. NP * (NP - 1) / 2) RETURN
      IFAULT = 0
C
      DO 100 J = 1, NP
      I = NP - J + 1
      BETA(I) = THETAB(I)
      IF (I .EQ. NP) GOTO 100
      NEXTR = (I - 1) * (NP + NP - I) / 2 + 1
      IP = I + 1
      DO 50 K = IP, NP
      BETA(I) = BETA(I) - RBAR(NEXTR) * BETA(K)
      NEXTR = NEXTR + 1
   50 CONTINUE
  100 CONTINUE
      RETURN
      END
C
      SUBROUTINE ANALYZ(MAXP, MAXR, NP, NRBAR, INPUT, OUTPUT, X, DD,
     $ THETA, R, IFAULT)
C
C         ALGORITHM AS 75.5   APPL. STATIST. (1974) VOL.23, P.448
C
C         THIS SUBROUTINE READS DATA FROM CHANNEL INPUT AND
C         PRODUCES LEAST SQUARES VARIANCE COMPONENT ANALYSES ON
C         CHANNEL OUTPUT. THE INPUT MEDIUM IS ASSUMED TO
C         CONTAIN FIRST A VALUE FOR THE INTEGER NP, THE NUMBER OF
C         INDEPENDENT VARIATES, AND THEN A SEQUENCE OF RECORDS
C         CONTAINING THE ROWS OF X AND Y TOGETHER WITH WEIGHTS.
C         THE NUMBER ZERO (READ AS A WEIGHT) WILL INDICATE THE END OF
C         EACH DATA SET, AND UPON READING THIS AN ANALYSIS WILL BE
C         PRODUCED. SUCCESSIVE ANALYSES WILL BE DONE UNTIL NP,
C         THE NUMBER OF INDEPENDENT VARIATES, IS READ AS ZERO.
C
C         NO NEED TO CHECK THE IFAULT PARAMETER AFTER CALLING
C         AUXILIARY ROUTINES AS PARAMETERS HAVE ALREADY BEEN
C         CHECKED HERE.
C
      INTEGER OUTPUT
      LOGICAL FIRST
      REAL X(MAXP), DD(MAXP), THETA(MAXP), R(MAXR), ERROR, W, Y, ZERO,
     $ ONE, EPS, TOL, ZABS
C
      DATA ZERO, ONE, EPS, TOL /0.0, 1.0, 1.0E-8, 1.0E-4/
C
      ZABS(Y) = ABS(Y)
C
C         INPUT FORMATS
C
    1 FORMAT(I3)
    2 FORMAT(F10.6)
    3 FORMAT(F10.6)
    4 FORMAT(I3)
    5 FORMAT(F10.6)
C
C         OUTPUT FORMATS
C
   11 FORMAT(1H1, I5, 18H OBSERVATIONS READ/ 19H DIAGONAL MATRIX IS/
     $ (1X, 5G15.8))
   12 FORMAT(21H CONFOUNDED CONTRASTS)
   13 FORMAT(1X, 5G15.8)
   14 FORMAT(29H SUM OF SQUARES DECOMPOSITION/ (1X, 5G15.8))
   15 FORMAT(22H SUM OF SQUARES ERROR , G15.8)
   16 FORMAT(24H REGRESSION COEFFICIENTS/ (1X, 5G15.8))
```

```
C
C          READ NP, CHECK ITS VALUE AND INITIALISE ARRAYS
C
      IFAULT = 0
   20 READ (INPUT, 1) NP
      IF (NP .EQ. 0) RETURN
      NRBAR = NP * (NP - 1) / 2
      IF (NP .LT. 0 .OR. NP .GT. MAXP .OR. NRBAR .GT. MAXR) GOTO 200
      DO 30 K = 1, NP
      DD(K) = ZERO
      THETA(K) = ZERO
   30 CONTINUE
      DO 40 K = 1, NRBAR
   40 R(K) = ZERO
      ERROR = ZERO
      N = 0
C
C          READ WEIGHT AND CHECK ITS VALUE
C
   50 READ (INPUT, 2) W
      IF (W .EQ. ZERO) GOTO 90
      IF (W .GT. ZERO) N = N + 1
      IF (W .LT. ZERO) N = N - 1
C
C          READ THE Y-VALUE AND ALL CORRESPONDING X-VALUES
C
      READ (INPUT, 3) Y
      DO 60 K = 1, NP
   60 X(K) = ZERO
   70 READ (INPUT, 4) K
      IF (K .LT. 0 .OR. K .GT. NP) GOTO 300
      IF (K .EQ. 0) GOTO 80
      READ (INPUT, 5) X(K)
      GOTO 70
   80 CALL INCLUD(NP, NRBAR, W, X, Y, DD, R, THETA, ERROR, IFAIL)
      GOTO 50
C
C          BEGIN OUTPUT
C
   90 WRITE (OUTPUT, 11) N, (DD(I), I = 1, NP)
      FIRST = .TRUE.
      DO 130 J = 1, NP
      IF (ZABS(DD(J)) .GE. EPS) GOTO 130
C
C          CONFOUNDING DISCOVERED
C
      IF (.NOT. FIRST) GOTO 100
      FIRST = .FALSE.
      WRITE (OUTPUT, 12)
  100 CALL CONFND(NP, NRBAR, J, R, X, IFAIL)
      WRITE (OUTPUT, 13) (X(I), I = 1, NP)
C
C          CHOOSE RESOLVING CONSTRAINT
C
      IF (J .EQ. 1) GOTO 120
      JM = J - 1
      DO 110 K = 1, JM
      IF (ZABS(X(K)) .LE. TOL) GOTO 110
      X(K) = ZERO
      GOTO 120
  110 CONTINUE
  120 CALL INCLUD(NP, NRBAR, ONE, X, ZERO, DD, R, THETA, ERROR, IFAIL)
  130 CONTINUE
      CALL SSDCMP(NP, DD, THETA, X, IFAIL)
      WRITE (OUTPUT, 14) (X(I), I = 1, NP)
      WRITE (OUTPUT, 15) ERROR
      CALL REGRSS(NP, NRBAR, R, THETA, X, IFAIL)
      WRITE (OUTPUT, 16) (X(I), I = 1, NP)
      GOTO 20
```

```
C
C          ERROR RETURNS
C
  200 IFAULT = 2
      RETURN
  300 IFAULT = 3
      RETURN
      END
```

Algorithm AS 76

AN INTEGRAL USEFUL IN CALCULATING
NON-CENTRAL t AND BIVARIATE NORMAL PROBABILITIES

By J. C. Young and Ch. E. Minder
Faculty of Mathematics,
The University of Waterloo,
Ontario, Canada, N2L 3G1

Keywords: Bivariate normal integral; Owen's T-function; Non-central t.

LANGUAGE

Fortran 66 and 77.

DESCRIPTION AND PURPOSE

The real function *TFN* calculates the function

$$T(h, a) = 1/(2\pi) \int_0^a \frac{\exp\{(-h^2/2)(1 + x^2)\}}{1 + x^2} \, dx \quad (-\infty < h, a < +\infty)$$

for given values h and a. This is the function used by Owen (1956) in setting up tables of the bivariate normal probability function and by Donnelly (1973) in an algorithm for its evaluation. It is the same function that is evaluated by *FUNCT* in algorithm AS 4 (Cooper, 1968a). *TFN* does not, however, require a normal probability integral as an auxiliary function. In addition *TFN* is accurate to at least six decimal places while AS 4 is sometimes inaccurate, irrespective of the value of its accuracy parameter *EPS*, especially for larger values of h and a.

The $T(h, a)$ function is used by Algorithm AS 5 (Cooper, 1968b) in evaluating the non-central t-distribution.

NUMERICAL METHOD

It is interesting to note that the above advantages were achieved simply by using Gaussian quadrature (see, for example, Stroud and Secrest, 1966, for details and tables). Although 10-point quadrature is used here, adequate accuracy for many applications can be achieved using fewer points, thus saving a proportionate amount of time. More points will, of course, lead to increased accuracy.

When the values of h and a are such that the integrand becomes relatively very small over a large portion of the range of integration, the program uses a truncation procedure to find an upper limit for the range of quadrature. This ensures accurate evaluation of that portion of the integrand that contributes to the relevant significant digits of the integral.

STRUCTURE

REAL FUNCTION TFN (X, FX)

Formal parameters

X	Real	input:	the value of h.
FX	Real	input:	the value of a.

Constants

NG	Integer	half the number of points used for Gaussian quadrature
U	Real array (NG)	half the tabulated abscissae used for Gaussian quadrature
R	Real array (NG)	half the tabulated weights used for Gaussian quadrature
TP	Real	$1.0/2\pi$
TV1	Real	square root of smallest number to be distinguished from zero
TV2	Real	for $x > TV2$, $\exp\{-x^2/2\}$ will be considered zero
TV3	Real	determines the truncation point. $\rho = \exp(TV3)$ is the ratio of the maximum value of the integrand in the expression for $T(h, a)$ (its value at $x = 0$) to its value at $x = $ 'truncation point'
TV4	Real	determines the precision with which the truncation point is to be found

TIME

The speeds of *TFN* and AS 4 depend in different ways on the values of h and a. AS 4 is slightly faster in some regions ($|a| < 1$ for some values of h) and *TFN* is much faster in others (part of the region $|a| < 1$ and most of the region $|a| > 1$). On the average, over the whole range of reasonable values of h and a, *TFN* will usually be noticeably faster than AS 4.

ACCURACY

Extensive checks with Owen's (1956) tables showed that *TFN* is accurate to at least six decimal places. In addition, comparison with two other routines computing the function $T(h, a)$ suggests accuracy to at least four significant digits for $T(h, a) \geqslant 10^{-20}$. Unfortunately the lack of tables of $T(h, a)$ precludes a more definite statement on the accuracy of *TFN*.

PRECISION

The routine can be converted into double precision by changing *REAL FUNCTION* to *DOUBLE PRECISION FUNCTION* and *REAL* to *DOUBLE*

PRECISION, replacing the real constants in the *DATA* statements by double precision values, and changing the right-hand sides of the five statement functions to *DABS, DEXP, DLOG, DSIGN* and *DATAN* respectively.

ACKNOWLEDGEMENT

We acknowledge the helpful comments and advice of the Algorithms Editor during the preparation of this algorithm.

EDITORS' REMARKS

This routine has been changed in accordance with Remarks AS R26 (Hill, 1978) and AS R30 (Thomas, 1979), and the corrections on pages 113 and 336 of *Applied Statistics,* **28** (1979). The only other changes made to this routine are those facilitating the change to double precision outlined above.

REFERENCES

Cooper, B. E. (1968a) Algorithm AS 4. An auxiliary function for distribution integrals. *Appl. Statist.,* **17,** 190–192.

Cooper, B. E. (1968b) Algorithm AS 5. The integral of the non-central *t*-distribution. *Appl. Statist.,* **17,** 193–194. (See also this book, page 40).

Donnelly, T. G. (1973) Algorithm 462. Bivariate normal distribution. *Commun. Ass. Comput. Mach.,* **16,** 638.

Hill, I. D. (1978) Remark AS R26. *Appl. Statist.,* **27,** 379.

Owen, D. B. (1956) Tables for computing bivariate normal probabilities. *Ann. Math. Statist.,* **27,** 1075–1090.

Stroud, H. A. and Secrest, D. (1966) *Gaussian Quadrature Formulas.* Englewood Cliffs, New Jersey: Prentice-Hall.

Thomas, G. E. (1979) Remark AS R30. *Appl. Statist.,* **28,** 113.

```
      REAL FUNCTION TFN(X, FX)
C
C         ALGORITHM AS 76  APPL. STATIST. (1974) VOL.23, P.455
C
C         CALCULATES THE T-FUNCTION OF OWEN, USING GAUSSIAN
C         QUADRATURE
C
      REAL U(5), R(5), X, FX, TP, TV1, TV2, TV3, TV4, ZERO,
     $   QUART, HALF, ONE, TWO, R1, R2, RT, XS, X1, X2, FXS,
     $   ZABS, ZEXP, ZLOG, ZSIGN, ZATAN
C
      DATA      U(1),       U(2),       U(3),       U(4),       U(5)
     $   /0.0744372, 0.2166977, 0.3397048, 0.4325317, 0.4869533/
C
      DATA      R(1),       R(2),       R(3),       R(4),       R(5)
     $   /0.1477621, 0.1346334, 0.1095432, 0.0747257, 0.0333357/
C
      DATA NG,       TP,     TV1, TV2, TV3,     TV4
     $    / 5, 0.159155, 1.0E-35, 15.0, 15.0, 1.0E-5/
```

```
C
      DATA ZERO, QUART, HALF, ONE, TWO
     $    / 0.0,  0.25,  0.5, 1.0, 2.0/
C
      ZABS(X) = ABS(X)
      ZEXP(X) = EXP(X)
      ZLOG(X) = ALOG(X)
      ZSIGN(X1, X2) = SIGN(X1, X2)
      ZATAN(X) = ATAN(X)
C
C        TEST FOR X NEAR ZERO
C
      IF (ZABS(X) .GE. TV1) GOTO 5
      TFN = TP * ZATAN(FX)
      RETURN
C
C        TEST FOR LARGE VALUES OF ABS(X)
C
   5 IF (ZABS(X) .GT. TV2) GOTO 10
C
C        TEST FOR FX NEAR ZERO
C
      IF (ZABS(FX) .GE. TV1) GOTO 15
  10 TFN = ZERO
      RETURN
C
C        TEST WHETHER ABS(FX) IS SO LARGE THAT IT MUST BE
C        TRUNCATED
C
  15 XS = -HALF * X * X
      X2 = ZABS(FX)
      FXS = FX * FX
      IF (ZLOG(ONE + FXS) - XS * FXS .LT. TV3) GOTO 25
C
C        COMPUTATION OF TRUNCATION POINT BY NEWTON ITERATION
C
      X1 = HALF * X2
      FXS = QUART * FXS
  20 RT = FXS + ONE
      X2 = X1 + (XS * FXS + TV3 - ZLOG(RT)) / (TWO * X1 *
     $  (ONE / RT - XS))
      FXS = X2 * X2
      IF (ZABS(X2 - X1) .LT. TV4) GOTO 25
      X1 = X2
      GOTO 20
C
C        GAUSSIAN QUADRATURE
C
  25 RT = ZERO
      DO 30 I = 1, NG
      R1 = ONE + FXS * (HALF + U(I)) ** 2
      R2 = ONE + FXS * (HALF - U(I)) ** 2
      RT = RT + R(I) * (ZEXP(XS * R1) / R1 + ZEXP(XS * R2) / R2)
  30 CONTINUE
      TFN = ZSIGN(RT * X2 * TP, FX)
      RETURN
      END
```

Algorithm AS 83

COMPLEX DISCRETE FAST FOURIER TRANSFORM
By Donald M. Monro
Engineering in Medicine Laboratory, Imperial College, London

Present address: Department of Electrical Engineering, Imperial College of Science and Technology, Exhibition Road, London SW7 2BT, UK.

Keywords: Fast Fourier transform; Discrete Fourier transform; Fourier series; Spectral analysis.

LANGUAGE

Fortran 66 and 77.

DESCRIPTION AND PURPOSE

The purpose of this algorithm is the efficient evaluation of the complex discrete Fourier transform (DFT) of a sequence of length N, $\mathbf{X} = \{X_0, X_1, \ldots, X_{N-1}\}$, defined for this procedure as

$$Y_n = \frac{1}{N} \sum_{k=0}^{N-1} X_k \exp(-j2\pi kn/N), \; n = 0, 1, \ldots, N-1, \qquad (1)$$

where $\mathbf{Y} = \{Y_0, Y_1, \ldots, Y_{N-1}\}$ is the transformed sequence. The inverse transform is

$$X_k = \sum_{n=0}^{N-1} Y_n \exp(j2\pi kn/N), \; k = 0, 1, \ldots, N-1. \qquad (2)$$

Widespread interest in the fast Fourier transform began with the description of the algorithm by Cooley and Tukey (1965), and a number of authors have discussed practical considerations related to its implementation, notably Gentleman and Sande (1966) and Singleton (1967). By reducing the DFT to a series of small transforms, typically of length 2 or 4, the fast Fourier transform eliminates many redundant operations and so achieves a remarkable improvement in speed. The penalty is that the sequence length must be capable of decomposition into small prime factors, which most often means that N must be a power of 2, as in this algorithm.

This evaluation uses the DFT subroutine *FASTG*, whose optimization takes into consideration the requirements of both large computers, in which indexing is likely to account for an appreciable proportion of execution time, and of small machines in which the speed will be dominated by the amount of floating point arithmetic required. The full complex DFT is normally evaluated by calling subroutine *FASTF,* which itself calls on subroutine *FASTG* to perform the transform and subroutine *SCRAM* for reordering of the results. In developing the routine, improvements in speed have been incorporated only where they do not introduce unacceptable increases in length. A report is available from the author describing in detail the factors contributing to this implementation (Monro, 1971).

FASTG is essentially an implementation of the Sande–Tukey algorithm. Various other forms have been tried, and two possible sources of variation are worthy of mention. The particular recursion used to generate sine and cosine values with correction for propagated errors is the one suggested by Singleton (1967). It has been found by comparison that the errors in the transform produced in this way are actually smaller than those resulting from direct use of the library sine and cosine functions in place of the recursion. The use of a table to hold these sine and cosine values could lead to time savings after the first transform of a fixed length. However, this would require at least $N/4 - 1$ extra real variables to be stored and would also add substantially to the coded length of *FASTG*. In view of the desire to keep storage requirements reasonable, a small compromise in speed has been accepted in this instance. In fact the first transform of a given length using a sine table necessarily involves calculation of the table and is actually slower using library functions than the transform with recursive sine generation.

The second type of variation is in the unscrambling of the transform results. Several different schemes have been evaluated, and the fastest version is presented here as *SUBROUTINE SCRAM*. It uses a large number of nested loops as first suggested by Gentleman and Sande (1966). This method accounts for about 5 per cent of the transform time but could be rejected by Fortran compilers with restrictions on the number of nested *DO*-loops. An alternative version simulates the nested loops using arrays for the loop indices, increments and limits but takes twice the time. Two other possibilities which compare in speed to the slower version are an interleaved unscrambling in which rearrangement is done after each transform stage, and a direct bit-reversal operation using *shift* and *and* functions which could offer speed and storage advantages in a machine code implementation.

STRUCTURE

SUBROUTINE FASTF (XREAL, XIMAG, ISIZE, ITYPE, IFAULT)

Formal parameters

XREAL	Real array (*ISIZE*)	input:	the real part of the original sequence.
		output:	the real part of the transform.

XIMAG	Real array (*ISIZE*)	input:	the imaginary part of the original sequence.
		output:	the imaginary part of the transform.
ISIZE	Integer	input:	the length of transform. Must be positive and a power of 2.
ITYPE	Integer	input:	the type of transform. A forward transform is found if *ITYPE* is positive, inverse if negative. If *ITYPE* is zero, no transform is done.
IFAULT	Integer	output:	an error indicator, equal to 1 if *ISIZE* is not a power of 2 or *ISIZE* < 4 or *ISIZE* > 2^{MAX2}; 0 otherwise.

Auxiliary algorithms

Subroutine *FASTF* calls subroutines *FASTG* (AS 83.2) and *SCRAM* (AS 83.3).

RESTRICTIONS

The transform length, given by *ISIZE*, must be a power of two and lie within the limits $4 \leqslant ISIZE \leqslant 2^{MAX2}$. This restriction is imposed by the unscrambling algorithm but is unlikely to be of practical importance since few computers at present have sufficient memory to attempt such a large transform. The length is checked by *FASTF* and if illegal the program returns immediately with the data untouched, as it does if *ITYPE* is zero. *MAX2* is set in a *DATA* statement, and as published takes the value 20.

TIME

The times tabulated below are typical runs with pseudorandom real sequences on the CDC 6400 computer, using the FTN compiler.

Transform length *N*	Forward transform time (sec)	Inverse transform time (sec)
64	0·018	0·017
128	0·035	0·034
256	0·074	0·074
512	0·165	0·162
1,024	0·357	0·343
2,048	0·772	0·757
4,096	1·63	1·64

ACCURACY

The accuracy of the transform has been tested by evaluating forward transforms followed by inverse transforms of various real sequences for a wide range of lengths N with the results then compared point by point with the original. The errors approach a consistent increase with $\log_2 N$ for large N, the actual magnitude of the error depending on the word length of the computer used and the magnitude of the original data sequence. For pseudorandom real data scaled between 0 and 1, one decimal digit is lost in the result at a moderate value of N, for example 1,024 on the CDC 6400, but extrapolation of the observed error would indicate that a second digit is not completely lost even at the longest transform currently allowed, 2^{20}.

PRECISION

The routines can be converted into double precision by changing *REAL* to *DOUBLE PRECISION* in each of the three routines, replacing the real constants in the *DATA* statements by double precision values, and changing the right-hand sides of the three statement functions in *FASTG* to their double precision equivalents (but note that in standard Fortran 66 there is no equivalent to *FLOAT*, though many compilers offer such a routine; alternatively *DBLE(FLOAT(K))* may be used).

ACKNOWLEDGEMENTS

Development of these programs was based largely on the facilities of the Imperial College Computer Centre, but also to a significant extent on facilities provided by the UK Medical Research Council for research into the analysis of biological signals by Professor B. McA. Sayers, whose interest in this work has been much appreciated.

EDITORS' REMARKS

This algorithm has been substantially amended, but without altering the basic structure of the original. These changes have been principally to bring it into line with the style of the other algorithms presented here, and to facilitate the change to double precision outlined above. The *IFAULT* parameter, and the test for the validity of *ISIZE*, were not originally included. Formerly included was an additional alternative version of *SCRAM* which did not require the twenty nested *DO*-loops. If a compiler will not accept the version of *SCRAM* presented here, users are referred to this alternative (Monro, 1975).

REFERENCES

Cooley, J. W. and Tukey, J. W. (1965) An algorithm for the machine calculation of complex Fourier series. *Math. Comp.*, **19**, 297–301.

Gentleman, W. M. and Sande, G. (1966) Fast Fourier transforms – for fun and profit. *AFIPS Proceedings of the Fall Joint Computer Conference*, **19**, 563–578.

Monro, D. M. (1971) *Implementing the fast Fourier transform.* Engineering in Medicine Laboratory, Research Report No. 4.

Monro, D. M. (1975) Algorithm AS 83. Complex discrete fast Fourier transform. *Appl. Statist.*, **24**, 153–160.

Singleton, R. C. (1967) On computing the fast Fourier transform. *Commun. Ass. Comput. Mach.*, **10**, 647–654.

```
      SUBROUTINE FASTF(XREAL, XIMAG, ISIZE, ITYPE, IFAULT)
C
C         ALGORITHM AS 83.1  APPL. STATIST. (1975) VOL.24, P.153
C
C         RADIX 4 COMPLEX DISCRETE FAST FOURIER TRANSFORM WITH
C         UNSCRAMBLING OF THE TRANSFORMED ARRAYS
C
      REAL XREAL(ISIZE), XIMAG(ISIZE)
C
C         CHECK FOR VALID TRANSFORM SIZE - UP TO 2 ** MAX2
C
      DATA MAX2 /20/
C
      II = 4
      DO 2 IPOW = 2, MAX2
      IF (II - ISIZE) 1, 4, 3
    1 II = II * 2
    2 CONTINUE
C
C         IF THIS POINT IS REACHED A SIZE ERROR HAS OCCURRED
C
    3 IFAULT = 1
      RETURN
    4 IFAULT = 0
      IF (ITYPE .EQ. 0) RETURN
C
C         CALL FASTG (ALGORITHM AS 83.2) TO PERFORM THE TRANSFORM
C
      CALL FASTG(XREAL, XIMAG, ISIZE, ITYPE)
C
C         CALL SCRAM (ALGORITHM AS 83.3)
C         TO UNSCRAMBLE THE RESULTS
C
      CALL SCRAM(XREAL, XIMAG, ISIZE, IPOW)
      RETURN
      END
C
      SUBROUTINE FASTG(XREAL, XIMAG, N, ITYPE)
C
C         ALGORITHM AS 83.2  APPL. STATIST. (1975) VOL.24, P.153
C
C         RADIX 4 COMPLEX DISCRETE FAST FOURIER TRANSFORM WITHOUT
C         UNSCRAMBLING, SUITABLE FOR CONVOLUTIONS OR OTHER
C         APPLICATIONS WHICH DO NOT REQUIRE UNSCRAMBLING.
C         SUBROUTINE FASTF USES THIS ROUTINE FOR TRANSFORMATION
C         AND ALSO PROVIDES UNSCRAMBLING
C
      REAL XREAL(N), XIMAG(N), BCOS, BSIN, CW1, CW2, CW3, PI,
     $ SW1, SW2, SW3, TEMPR, X1, X2, X3, XS0, XS1, XS2, XS3,
     $ Y1, Y2, Y3, YS0, YS1, YS2, YS3, Z, ZERO, HALF, ONE,
     $ ONE5, TWO, FOUR, ZATAN, ZFLOAT, ZSIN
```

```
C
      DATA ZERO, HALF, ONE, ONE5, TWO, FOUR
     $      /0.0,  0.5,  1.0,  1.5,  2.0,  4.0/
C
      ZATAN(Z) = ATAN(Z)
      ZFLOAT(K) = FLOAT(K)
      ZSIN(Z) = SIN(Z)
C
      PI = FOUR * ZATAN(ONE)
      IFACA = N / 4
      IF (ITYPE .GT. 0) GOTO 5
C
C         IF THIS IS TO BE AN INVERSE TRANSFORM, CONJUGATE THE DATA
C
      DO 4 K = 1, N
    4 XIMAG(K) = -XIMAG(K)
    5 IFCAB = IFACA * 4
C
C         DO THE TRANSFORMS REQUIRED BY THIS STAGE
C
      Z = PI / ZFLOAT(IFCAB)
      BCOS = -TWO * ZSIN(Z) ** 2
      BSIN = ZSIN(TWO * Z)
      CW1 = ONE
      SW1 = ZERO
      DO 10 LITLA = 1, IFACA
      DO 8 IO = LITLA, N, IFCAB
C
C         THIS IS THE MAIN CALCULATION OF RADIX 4 TRANSFORMS
C
      I1 = IO + IFACA
      I2 = I1 + IFACA
      I3 = I2 + IFACA
      XS0 = XREAL(IO) + XREAL(I2)
      XS1 = XREAL(IO) - XREAL(I2)
      YS0 = XIMAG(IO) + XIMAG(I2)
      YS1 = XIMAG(IO) - XIMAG(I2)
      XS2 = XREAL(I1) + XREAL(I3)
      XS3 = XREAL(I1) - XREAL(I3)
      YS2 = XIMAG(I1) + XIMAG(I3)
      YS3 = XIMAG(I1) - XIMAG(I3)
      XREAL(IO) = XS0 + XS2
      XIMAG(IO) = YS0 + YS2
      X1 = XS1 + YS3
      Y1 = YS1 - XS3
      X2 = XS0 - XS2
      Y2 = YS0 - YS2
      X3 = XS1 - YS3
      Y3 = YS1 + XS3
      IF (LITLA .GT. 1) GOTO 7
      XREAL(I2) = X1
      XIMAG(I2) = Y1
      XREAL(I1) = X2
      XIMAG(I1) = Y2
      XREAL(I3) = X3
      XIMAG(I3) = Y3
      GOTO 8
C
C         MULTIPLY BY TWIDDLE FACTORS IF REQUIRED
C
    7 XREAL(I2) = X1 * CW1 + Y1 * SW1
      XIMAG(I2) = Y1 * CW1 - X1 * SW1
      XREAL(I1) = X2 * CW2 + Y2 * SW2
      XIMAG(I1) = Y2 * CW2 - X2 * SW2
      XREAL(I3) = X3 * CW3 + Y3 * SW3
      XIMAG(I3) = Y3 * CW3 - X3 * SW3
    8 CONTINUE
      IF (LITLA .EQ. IFACA) GOTO 10
```

```
C
C          CALCULATE A NEW SET OF TWIDDLE FACTORS
C
      Z = CW1 * BCOS - SW1 * BSIN + CW1
      SW1 = BCOS * SW1 + BSIN * CW1 + SW1
      TEMPR = ONE5 - HALF * (Z * Z + SW1 * SW1)
      CW1 = Z * TEMPR
      SW1 = SW1 * TEMPR
      CW2 = CW1 * CW1 - SW1 * SW1
      SW2 = TWO * CW1 * SW1
      CW3 = CW1 * CW2 - SW1 * SW2
      SW3 = CW1 * SW2 + CW2 * SW1
   10 CONTINUE
      IF (IFACA .LE. 1) GOTO 14
C
C          SET UP THE TRANSFORM SPLIT FOR THE NEXT STAGE
C
      IFACA = IFACA / 4
      IF (IFACA .GT. 0) GOTO 5
C
C          THIS IS THE CALCULATION OF A RADIX TWO STAGE
C
      DO 13 K = 1, N, 2
      TEMPR = XREAL(K) + XREAL(K + 1)
      XREAL(K + 1) = XREAL(K) - XREAL(K + 1)
      XREAL(K) = TEMPR
      TEMPR = XIMAG(K) + XIMAG(K + 1)
      XIMAG(K + 1) = XIMAG(K) - XIMAG(K + 1)
      XIMAG(K) = TEMPR
   13 CONTINUE
   14 IF (ITYPE .GT. 0) GOTO 17
C
C          IF THIS WAS AN INVERSE TRANSFORM, CONJUGATE THE RESULT
C
      DO 16 K = 1, N
   16 XIMAG(K) = -XIMAG(K)
      RETURN
C
C          IF THIS WAS A FORWARD TRANSFORM, SCALE THE RESULT
C
   17 Z = ONE / ZFLOAT(N)
      DO 18 K = 1, N
      XREAL(K) = XREAL(K) * Z
      XIMAG(K) = XIMAG(K) * Z
   18 CONTINUE
      RETURN
      END
C
      SUBROUTINE SCRAM(XREAL, XIMAG, N, IPOW)
C
C          ALGORITHM AS 83.3   APPL. STATIST. (1975) VOL.24, P.153
C
C          SUBROUTINE FOR UNSCRAMBLING FFT DATA.
C
      REAL XREAL(N), XIMAG(N), TEMPR
      INTEGER L(19)
      EQUIVALENCE      (L1,    L(1)), (L2,    L(2)), (L3,    L(3)),
     $  (L4,    L(4)), (L5,    L(5)), (L6,    L(6)), (L7,    L(7)),
     $  (L8,    L(8)), (L9,    L(9)), (L10, L(10)), (L11, L(11)),
     $  (L12, L(12)), (L13, L(13)), (L14, L(14)), (L15, L(15)),
     $  (L16, L(16)), (L17, L(17)), (L18, L(18)), (L19, L(19))
C
      II = 1
      ITOP = 2 ** (IPOW - 1)
      I = 20 - IPOW
      DO 5 K = 1, I
    5 L(K) = II
      LO = II
```

```
        I = I + 1
        DO 6 K = I, 19
        II = II * 2
        L(K) = II
 6   CONTINUE
        II = 0
        DO 9 J1 = 1, L1, L0
        DO 9 J2 = J1, L2, L1
        DO 9 J3 = J2, L3, L2
        DO 9 J4 = J3, L4, L3
        DO 9 J5 = J4, L5, L4
        DO 9 J6 = J5, L6, L5
        DO 9 J7 = J6, L7, L6
        DO 9 J8 = J7, L8, L7
        DO 9 J9 = J8, L9, L8
        DO 9 J10 = J9, L10, L9
        DO 9 J11 = J10, L11, L10
        DO 9 J12 = J11, L12, L11
        DO 9 J13 = J12, L13, L12
        DO 9 J14 = J13, L14, L13
        DO 9 J15 = J14, L15, L14
        DO 9 J16 = J15, L16, L15
        DO 9 J17 = J16, L17, L16
        DO 9 J18 = J17, L18, L17
        DO 9 J19 = J18, L19, L18
        J20 = J19
        DO 9 I = 1, 2
        II = II + 1
        IF (II .GE. J20) GOTO 8
C
C           J20 IS THE BIT-REVERSE OF II
C           PAIRWISE INTERCHANGE
C
        TEMPR = XREAL(II)
        XREAL(II) = XREAL(J20)
        XREAL(J20) = TEMPR
        TEMPR = XIMAG(II)
        XIMAG(II) = XIMAG(J20)
        XIMAG(J20) = TEMPR
 8   J20 = J20 + ITOP
 9   CONTINUE
        RETURN
        END
```

Algorithm AS 91

THE PERCENTAGE POINTS OF THE χ^2 DISTRIBUTION

By D. J. Best and D. E. Roberts[†]

Division of Mathematics and Statistics, CSIRO,
P.O. Box 52, North Ryde, NSW 2113, Australia

Keywords: Chisquared distribution; Percentage points; Taylor series.

LANGUAGE

Fortran 66 and 77.

DESCRIPTION AND PURPOSE

Given a value P of the lower tail area of the χ^2 distribution with ν degrees of freedom, the subroutine computes the corresponding point z. Thus

$$P = \int_0^z \phi(u)\, du, \tag{1}$$

where

$$\phi(u) = 2^{-\frac{1}{2}\nu}\, \{\Gamma(\tfrac{1}{2}\nu)\}^{-1}\, \exp(-\tfrac{1}{2}u)u^{\frac{1}{2}\nu-1}, \ \nu > 0.$$

The subroutine is written so that z may be calculated (for ν not necessarily integral) as exactly as the user's computer allows. Thus our subroutine is both more general and more accurate than ACM Algorithm 451 (Goldstein, 1973).

NUMERICAL METHOD

z is found from the Taylor series expansion (Hill and Davis, 1968)

$$z = z_0 + \sum_r c_r(z_0)\, \{E/\phi(z_0)\}^r(r!)^{-1}, \tag{2}$$

where z_0 is a suitable starting approximation,

$$c_1(u) = 1, \ c_{r+1}(u) = (r\psi + d/du)\, c_r(u),$$

$$E = P - \int_0^{z_0} \phi(u)\, du \ \text{and} \ \psi = \tfrac{1}{2} - (\tfrac{1}{2}\nu - 1)\, u^{-1}.$$

For many P, ν values the Wilson–Hilferty approximation (Kendall and Stuart, p. 372, 1969) can be used for z_0, viz.

$$z_{01} = \nu\{x(2/9\nu)^{\frac{1}{2}} + 1 - (2/9\nu)\}^3,$$

[†]*Present address:* Physics Department, Nuclear Development Corporation of South Africa (Pty) Ltd, Private Bag X256, Pretoria, South Africa.

where x is the lower $100P\%$ point of the standard Normal distribution. However, better starting approximations are necessary in the three limiting cases, $P \to 0$, $P \to 1$ and $\nu \to 0$.

(i) $P \to 0$ (small z): Equation (1) can be simplified to give

$$z_{02} = \{P\nu 2^{\frac{1}{2}\nu - 1} \, \Gamma(\tfrac{1}{2}\nu)\}^{2/\nu}.$$

z_{02} is better than z_{01} for $\nu < -1\cdot24 \ln P$. This criterion ensures that replacing $\exp(-\frac{1}{2}u)$ by 1 in (1) is in error by less than 10 per cent for $\nu \to 0$. For the special case $z_{02} < 2 \times 10^{-6}$, $z = z_{02}$ gives at least six significant figure accuracy.

(ii) $P \to 1$ (large z): Equation (1) can be simplified to give

$$z \doteq -2[\ln(1-P) - (\tfrac{1}{2}\nu - 1)\ln(\tfrac{1}{2}z) + \ln\{\Gamma(\tfrac{1}{2}\nu)\}].$$

For $z_{01} > 2\cdot2\nu + 6$ a better starting approximation than z_{01} is found to be

$$z_{03} = -2[\ln(1-P) - (\tfrac{1}{2}\nu - 1)\ln(\tfrac{1}{2}z_{01}) + \ln\{\Gamma(\tfrac{1}{2}\nu)\}].$$

(iii) $\nu \to 0$: For the special case $\nu \leqslant 0\cdot32$, P is expressed in terms of an approximation (Hastings, 1955) to the exponential integral and z_{04} found by Newton–Raphson iteration.

STRUCTURE

REAL FUNCTION PPCHI2 (P, V, G, IFAULT)

Formal parameters

P	Real	input:	value of lower tail area.
V	Real	input:	degrees of freedom para-meter.
G	Real	input:	the natural logarithm of $\Gamma(\tfrac{1}{2}\nu)$.
IFAULT	Integer	output:	a fault indicator, equal to: 1 if $P > 0\cdot999998$ or $P < 0\cdot000002$; 2 if $\nu \leqslant 0\cdot0$; 3 if the fault indicator of *FUNCTION GAMMDS* is greater than zero; 0 otherwise.

If a fault is detected *PPCHI2* is set equal to $-1\cdot0$.

Auxiliary algorithms
The following auxiliary subroutines are called:

REAL FUNCTION GAMMDS (Y, P, IFAULT) – Algorithm AS 147 (Lau, 1980) and *REAL FUNCTION PPND (P, IFAULT)* — Algorithm AS 111 (Beasley and Springer, 1977).

For the natural logarithm of $\Gamma(\frac{1}{2}\nu)$ any standard algorithm, such as ACM Algorithm 291 (Pike and Hill, 1966), may be used.

ACCURACY

If the appropriate starting approximation is used with seven terms in (2) only one evaluation is necessary to give at least six significant figures except for the small region $5 \cdot 6 < \nu < -4 \cdot 07 \ln(P) + 12 \cdot 21$ where two evaluations are necessary.

If more than six significant figures are required the *DATA* statement should be changed to alter E appropriately. When this is done more iterations of the first seven terms of (2) are performed as necessary.

PRECISION

The routine can be converted into double precision by changing *REAL FUNCTION* to *DOUBLE PRECISION FUNCTION* and *REAL* to *DOUBLE PRECISION*, replacing the real constants in the *DATA* statement by double precision values, and changing the right-hand sides of the four statement functions to the corresponding double precision functions, namely *DABS, DEXP, DLOG,* and *DSQRT* respectively.

EDITORS' REMARKS

This routine has been modified to use **AS 111** and **AS 147** as auxiliary routines in place of **AS 70** and **AS 32** respectively. The only other changes made are those facilitating the change to double precision outlined above.

REFERENCES

Beasley, J. D. and Springer, S. G. (1977) Algorithm AS 111. The percentage points of the normal distribution. *Appl. Statist.,* **26,** 118–121. (See also this book, page 188)

Goldstein, R. B. (1973) Algorithm 451: Chi-square quantiles. *Commun. Ass. Comput. Mach.,* **16,** 483–485.

Hastings, C., Jr (1955) *Approximations for Digital Computers.* Princeton: Princeton University Press.

Hill, G. W. and Davis, A. W. (1968) Generalized asymptotic expansions of Cornish–Fisher type. *Ann. Math. Statist.,* **39,** 1264–1273.

Kendall, M. G. and Stuart, A. (1969) *The Advanced Theory of Statistics,* Vol. 1. London: Griffin.

Lau, C.–L. (1980) Algorithm AS 147. A simple series for the incomplete gamma integral. *Appl. Statist.,* **29,** 113–114. (See also this book, page 203)

Pike, M. C. and Hill, I. D. (1966) Algorithm 291: Logarithm of the gamma function. *Commun. Ass. Comput. Mach.,* **9,** 694. (See also this book, page 243)

```
      REAL FUNCTION PPCHI2(P, V, G, IFAULT)
C
C         ALGORITHM AS 91   APPL. STATIST. (1975) VOL.24, P.385
C
C         TO EVALUATE THE PERCENTAGE POINTS OF THE CHI-SQUARED
C         PROBABILITY DISTRIBUTION FUNCTION.
C         P MUST LIE IN THE RANGE 0.000002 TO 0.999998,
C         V MUST BE POSITIVE,
C         G MUST BE SUPPLIED AND SHOULD BE EQUAL TO
C            LN(GAMMA(V/2.0))
C
      REAL P, V, G, GAMMDS, PPND, AA, E, ZERO, HALF, ONE, TWO,
     $ THREE, SIX, PMIN, PMAX, C1, C2, C3, C4, C5, C6, C7, C8,
     $ C9, C10, C11, C12, C13, C14, C15, C16, C17, C18, C19, C20,
     $ C21, C22, C23, C24, C25, C26, C27, C28, C29, C30, C31,
     $ C32, C33, C34, C35, C36, C37, C38, A, B, C, CH, P1, P2,
     $ Q, S1, S2, S3, S4, S5, S6, T, X, XX,
     $ ZABS, ZEXP, ZLOG, ZSQRT
C
      DATA            AA,        E,      PMIN,      PMAX
     $ /0.6931471806, 0.5E-6, 0.000002, 0.999998/
      DATA ZERO, HALF, ONE, TWO, THREE, SIX
     $ /0.0,   0.5,  1.0,  2.0,   3.0,  6.0/
      DATA        C1,        C2,        C3,        C4,        C5,        C6,
     $            C7,        C8,        C9,       C10,       C11,       C12,
     $           C13,       C14,       C15,       C16,       C17,       C18,
     $           C19,       C20,       C21,       C22,       C23,       C24,
     $           C25,       C26,       C27,       C28,       C29,       C30,
     $           C31,       C32,       C33,       C34,       C35,       C36,
     $           C37,       C38/
     $          0.01, 0.222222,      0.32,       0.4,      1.24,       2.2,
     $          4.67,      6.66,      6.73,     13.32,      60.0,      70.0,
     $          84.0,     105.0,     120.0,     127.0,     140.0,     175.0,
     $         210.0,     252.0,     264.0,     294.0,     346.0,     420.0,
     $         462.0,     606.0,     672.0,     707.0,     735.0,     889.0,
     $         932.0,     966.0,    1141.0,    1182.0,    1278.0,    1740.0,
     $        2520.0,    5040.0/
C
      ZABS(X) = ABS(X)
      ZEXP(X) = EXP(X)
      ZLOG(X) = ALOG(X)
      ZSQRT(X) = SQRT(X)
C
C         TEST ARGUMENTS AND INITIALIZE
C
      PPCHI2 = -ONE
      IFAULT = 1
      IF (P .LT. PMIN .OR. P .GT. PMAX) RETURN
      IFAULT = 2
      IF (V .LE. ZERO) RETURN
      IFAULT = 0
      XX = HALF * V
      C = XX - ONE
C
C         STARTING APPROXIMATION FOR SMALL CHI-SQUARED
C
      IF (V .GE. -C5 * ZLOG(P)) GOTO 1
      CH = (P * XX * ZEXP(G + XX * AA)) ** (ONE / XX)
      IF (CH .LT. E) GOTO 6
      GOTO 4
C
C         STARTING APPROXIMATION FOR V LESS THAN OR EQUAL TO 0.32
C
    1 IF (V .GT. C3) GOTO 3
      CH = C4
      A = ZLOG(ONE - P)
    2 Q = CH
      P1 = ONE + CH * (C7 + CH)
      P2 = CH * (C9 + CH * (C8 + CH))
```

```
        T = -HALF + (C7 + TWO * CH) / P1 - (C9 + CH * (C10 +
     $   THREE * CH)) / P2
        CH = CH - (ONE - ZEXP(A + G + HALF * CH + C * AA) *
     $   P2 / P1) / T
        IF (ZABS(Q / CH - ONE) .GT. C1) GOTO 2
        GOTO 4
C
C
C           CALL TO ALGORITHM AS 111 - NOTE THAT P HAS BEEN TESTED
C           ABOVE
C
      3 X = PPND(P, IF1)
C
C           STARTING APPROXIMATION USING WILSON AND HILFERTY ESTIMATE
C
        P1 = C2 / V
        CH = V * (X * ZSQRT(P1) + ONE - P1) ** 3
C
C           STARTING APPROXIMATION FOR P TENDING TO 1
C
        IF (CH .GT. C6 * V + SIX)
     $   CH = -TWO * (ZLOG(ONE - P) - C * ZLOG(HALF * CH) + G)
C
C           CALL TO ALGORITHM AS 147 AND CALCULATION OF SEVEN TERM
C           TAYLOR SERIES
C
      4 Q = CH
        P1 = HALF * CH
        P2 = P - GAMMDS(P1, XX, IF1)
        IF (IF1 .EQ. 0) GOTO 5
        IFAULT = 3
        RETURN
      5 T = P2 * ZEXP(XX * AA + G + P1 - C * ZLOG(CH))
        B = T / CH
        A = HALF * T - B * C
        S1 = (C19 + A * (C17 + A * (C14 + A * (C13 + A * (C12 +
     $   C11 * A))))) / C24
        S2 = (C24 + A * (C29 + A * (C32 + A * (C33 + C35 *
     $   A)))) / C37
        S3 = (C19 + A * (C25 + A * (C28 + C31 * A))) / C37
        S4 = (C20 + A * (C27 + C34 * A) + C * (C22 + A * (C30 +
     $   C36 * A))) / C38
        S5 = (C13 + C21 * A + C * (C18 + C26 * A)) / C37
        S6 = (C15 + C * (C23 + C16 * C)) / C38
        CH = CH + T * (ONE + HALF * T * S1 - B * C * (S1 - B *
     $   (S2 - B * (S3 - B * (S4 - B * (S5 - B * S6))))))
        IF (ZABS(Q / CH - ONE) .GT. E) GOTO 4
C
      6 PPCHI2 = CH
        RETURN
        END
```

Algorithm AS 97

REAL DISCRETE FAST FOURIER TRANSFORM

By Donald M. Monro

Engineering in Medicine Laboratory, Imperial College, London, UK

Present address: Department of Electrical Engineering, Imperial College of Science and Technology, Exhibition Road, London SW7 2BT, UK.

Keywords: Fast Fourier transform; Fourier series; Spectral analysis.

LANGUAGE

Fortran 66 and 77.

DESCRIPTION AND PURPOSE

This algorithm provides an efficient means of evaluating the discrete Fourier transform of sequences which are known to be real. Algorithm AS 83 (Monro, 1975) provided for the efficient evaluation of the complex discrete Fourier transform (DFT), in which for a sequence $\{X\}$ whose length M is a power of 2,

$$\{X\} = \{X_0, X_1, \ldots, X_{M-1}\},$$

the full complex transformation is defined as

$$Y_m = \frac{1}{M} \sum_{k=0}^{M-1} X_k\, e^{-j2\pi km/M}, \quad m = 0, 1, \ldots, M-1, \tag{1}$$

where $\{Y\} = \{Y_0, Y_1, \ldots, Y_{M-1}\}$ is the transformed sequence. $\{X\}$ can be recovered from the inverse transform,

$$X_k = \sum_{m=0}^{M-1} Y_m e^{j2\pi km/M}, \quad k = 0, 1, \ldots, M-1. \tag{2}$$

In the situation where $\{X\}$ consists only of real numbers, then a symmetry exists in the transform $\{Y\}$ which is easily shown to be $Y_{M-m} = Y_m^*$, $m = 0, 1, \ldots, M-1$.

This suggests that all the components above $Y_{M/2}$ are redundant, and that the transformation itself could be made more efficient for real sequences. Indeed the general approach has been known for, some time (Gentleman and Sande, 1966) although not fully described.

From equation (1) the unique part of the DFT can be re-written as

$$Y_m = \frac{1}{M} \sum_{l=0}^{M/2-1} X_{2l} e^{-j4\pi lm/M} + \frac{1}{M} e^{-j2\pi m/M} \sum_{l=0}^{M/2-1} X_{2l+1}\, e^{-j4\pi lm/M},$$

$$m = 0, 1, \ldots, M/2, \tag{3}$$

which expresses the DFT in terms of transforms of the even and odd numbered members of X taken separately; this does not itself lead to any saving in storage. If, instead, a complex transform of length $M/2$ is calculated using the even numbered members as a real part and odd numbered ones as an imaginary part, the result is

$$Z_m = \frac{2}{M} \sum_{l=0}^{M/2-1} X_{2l} e^{-j4\pi lm/M} + j \frac{2}{M} \sum_{l=0}^{M/2-1} X_{2l+1} e^{-j4\pi lm/M},$$

$$m = 0, 1, \ldots, M/2-1, \quad (4)$$

which is related to Y_m simply enough that one can be turned into the other according to the relation

$$4Y_m = Z_m + Z^*_{-m} - j e^{-j2\pi m/M} \{Z_m - Z^*_{-m}\}, \quad m = 0, 1, \ldots, M/2, \quad (5)$$

bearing in mind that $Z_{M/2}$ is the same as Z_0.

Subroutines *FORRT* and *REVRT* accomplish this process, using half the storage for array variables compared with the subroutine *FASTF*, algorithm AS 83 (Monro, 1975), and also offer substantial savings in time for longer transforms. The complex transform *FASTG* is used which was part of AS 83 and a modified scrambling routine *SCRAG* is used to accomplish the re-ordering into even and odd parts.

The re-ordering procedure is a simplification of earlier methods (Singleton, 1967). Ideally the separation of the even and odd parts should be done in a single array (otherwise no storage is saved) but this cannot be done as a single permutation. However, two stages of bit-reverse unscrambling can accomplish an in-place re-ordering. Suppose in a sequence of length $M = 2^N$ an address K is represented by its binary digits,

$$K = \{b_{N-1} b_{N-2} \ldots b_2 b_1 b_0\}.$$

Then to use the same storage for the complex transform, the even and odd components are to be separated into different halves of the array. If the odd members are to be moved to the top half, then the new address derived from the address K of any member becomes

$$\{b_0 b_{N-1} b_{N-2} \ldots b_2 b_1\},$$

i.e. a rotation of the binary address.

The necessary transfers do not occur in pairs. However, a bit-reversal operation does lead to pairwise exchanges of address, and it is shown easily that if the entire array is first bit-reverse scrambled and then the half arrays are scrambled separately, then the desired re-ordering is complete:

original	$\{b_{N-1} b_{N-2} \ldots b_2 b_1 b_0\}$
first permutation bit reverse within array length M	$\{b_0 b_1 b_2 \ldots b_{N-2} b_{N-1}\}$
second permutation bit reverse within array lengths $M/2$	$\{b_0 b_{N-1} b_{N-2} \ldots b_2 b_1\}.$

Bearing in mind that *FASTG* itself operates on data in normal order and must be followed by unscrambling, the full forward transform program *FORRT*, consists of the steps

(i)   Full bit-reversal (modulo *M*)
(ii) Half bit-reversal (modulo *M*/2)
(iii) Complex transformation
(iv) Half bit-reversal (modulo *M*/2)
(v)   Unravelling of the result (Equation (5)).

In the inverse procedure, the complex transform conveniently provides the half reversal, so that the inverse program *REVRT* has steps:

(i)   Combine the transform (the inverse of Equation (5))
(ii) Complex inverse transformation
(iii) Full bit-reversal (modulo *M*).

Subroutine *SCRAG* is a modification of the subroutine *SCRAM*, algorithm AS 83.3 (Monro, 1975). It performs bit-reverse re-ordering of a single array, and is used five times by *FORRT* and once by *REVRT*. This version uses nested *DO*-loops, but compiler restrictions could force the nest to be simulated using an alternative form based on AS 83.4 (Monro, 1975). For improvement in speed in *FORRT* two calls to subroutine *SCRAM* from AS 83 could replace four calls to *SCRAG* as indicated in the program. It is obvious that the post-scrambling *FASTG* is ideally suited to inverse transformation because of the few steps involved. Implementations of the Sande–Tukey algorithm are best made post-scrambling and so ideally suited for *REVRT*. Alternative implementations with pre-scrambling which would suit *FORRT* are best made using the Cooley–Tukey algorithm; in other words, separate complex transforms for *FORRT* and *REVRT* would be desirable strictly from a speed point of view. However, the time difference between *FORRT* and *REVRT* would hardly justify the extra program storage involved.

STRUCTURE

SUBROUTINE FORRT (X, M, IFAULT)

Formal parameters

X	Real array (*M*)	input:	the original real sequence $\{X\}$ of length *M*.
		output:	the complex result $\{Y\}$ stored as follows: the real parts of Y_0 to $Y_{M/2}$ in equation (1) are returned in the expected positions, locations $X(1)$ to $X(M/2+1)$.

There is no imaginary part for $m = 0$ or $m = M/2$ and the other imaginary parts are returned $M/2$ places above the corresponding real parts.

M	Integer	input:	the length of the transform which must be positive and a power of 2.
$IFAULT$	Integer	output:	an error indicator, equal to 1 if M is not a power of 2, or $M < 8$ or $M > 2^{MAX2}$; 0 otherwise.

SUBROUTINE REVRT (X, M, IFAULT)

Formal parameters

X	Real array (M)	input:	the complex transform $\{Y\}$ of length M arranged as output from *FORRT*.
		output:	a real untransformed sequence $\{X\}$.
M	Integer	input:	the length of the transform which must be positive and a power of 2.
$IFAULT$	Integer	output:	an error indicator, equal to 1 if M is not a power of 2, or $M < 8$ or $M > 2^{MAX2}$; 0 otherwise.

AUXILIARY ALGORITHMS

Subroutines *FASTG* and *SCRAG* are required. The former is Algorithm AS 83.2, and the latter a modified version of AS 83.3 (Monro, 1975), or AS 83.4 if necessary (see Editors' Remarks below), modified in either case by:

(1) deleting *XIMAG* from the parameter list;
(2) deleting *XIMAG(N)* from the declarations;
(3) deleting the lines
  $TEMPR = XIMAG(II)$
  $XIMAG(II) = XIMAG(J20)$
  $XIMAG(J20) = TEMPR$

RESTRICTIONS

The transform length, given by M for both *FORRT* and *REVRT*, must be a power of 2 and lie within the limits $8 \leqslant M \leqslant 2^{MAX2}$. (The restriction is imposed by the scrambling procedure.) This size is checked and if illegal both programs return with the data untouched, and *IFAULT* set to 1. *MAX2* is set in *DATA* statements in both *FORRT* and *REVRT,* and as published takes the value 21.

TIME

Table 1 gives comparative timings for the complex transform *FASTF,* and the real transforms *FORRT* and *REVRT* for an IBM 1800 computer under the TSX system using non-re-entrant library routines. The Fortran compiler on this machine does not accept variable dimensions so to make these subroutines compilable all the arrays are redimensioned as length 1; no other changes are required.

Table 1 — Timings in seconds on the IBM 1800

Transform length	Forward complex transform (*FASTF*)	Inverse complex transform (*FASTF*)	Forward real transform (*FORRT*)	Inverse real transform (*REVRT*)
64	1·43	1·41	1·09	0·92
128	3·28	3·26	2·29	1·98
256	7·19	7·16	4·97	4·40
512	16·15	16·11	10·52	9·41
1024	34·80	34·68	22·77	20·61

In Table 2 are found similar comparative timings for the CDC 6400 computer using the FTN compiler. Here, on a machine with much faster arithmetic the overheads involved in the ordering play a larger part, so that the improvement in speed is not so dramatic, particularly for *FORRT,* and the differences between *FORRT* and *REVRT* due to the re-ordering procedures are more marked.

Table 2 — Typical timings in seconds on the CDC 6400. The computer is time-sharing and these times are subject to considerable fluctuation due to overheads in swapping

Transform length	Forward complex transform (*FASTF*)	Inverse complex transform (*FASTF*)	Forward real transform (*FORRT*)	Inverse real transform (*REVRT*)
128	0·035	0·034	0·034	0·023
256	0·074	0·074	0·066	0·050
512	0·165	0·162	0·141	0·104
1024	0·357	0·343	0·290	0·215
2048	0·772	0·757	0·600	0·454
4096	1·63	1·64	1·25	0·957

ACCURACY

The accuracy of the transformation has been tested by evaluating forward transforms followed by inverse transforms for a wide range of lengths M, with the results then compared point by point with the original. The errors approach a consistent increase with $\log_2 M$ for large M, the actual magnitude of error depending on the word length of the computer used and the magnitude of the original data. For pseudorandom real data scaled between 0 and 1, one decimal digit is lost in the result at a moderate value of M, but extrapolation of the observed errors would indicate that a second digit is not completely lost even at the longest transform currently allowed, 2^{21}.

PRECISION

The routines can be converted into double precision by changing *REAL* to *DOUBLE PRECISION* in each of the two routines, replacing the real constants in the *DATA* statements by double precision values, and changing the right-hand sides of the statement functions to their double precision equivalents (but note that in standard Fortran 66 there is no equivalent to *FLOAT,* though many compilers offer such a routine; alternatively *DBLE(FLOAT(K))* may be used).

ACKNOWLEDGEMENT

Development of these programs was based largely on the facilities of the Imperial College Computer Centre, but also to a significant extent on facilities provided by the UK Medical Research Council for research into the analysis of biological signals by Professor B. McA. Sayers, whose interest in the work has been much appreciated.

EDITORS' REMARKS

This algorithm has been substantially amended, but without altering the basic structure of the original. These changes have been principally to bring it into line with the style of the other algorithms presented here, and to facilitate the change to double precision outlined above.

The reader is referred to AS 83 earlier for details of alternatives if a compiler limits the number of nested *DO*-loops.

The *IFAULT* parameter, testing the input value of M, has been added.

REFERENCES

Gentleman, W. M. and Sande, G. (1966) Fast Fourier transforms — for fun and profit. *AFIPS Proceedings of the Fall Joint Computer Conference,* **19,** 563–578.

Monro, D. M. (1975) Algorithm AS 83. Complex discrete fast Fourier transform. *Appl. Statist.,* **24,** 153–160. (See also this book, page 149)

Singleton, R. C. (1967) On computing the fast Fourier transform. *Commun. Ass. Comput. Mach.,* **10,** 647–654.

```
      SUBROUTINE FORRT(X, M, IFAULT)
C
C         ALGORITHM AS 97.1   APPL. STATIST. (1976) VOL.25, P.166
C
C         FORWARD DISCRETE FOURIER TRANSFORM IN ONE DIMENSION OF
C         REAL DATA USING COMPLEX TRANSFORM SUBROUTINE FASTG
C
C         X = ARRAY OF REAL INPUT DATA, TYPE REAL, DIMENSION M.
C         M = LENGTH OF THE TRANSFORM, MUST BE A POWER OF 2.
C         THE MINIMUM LENGTH IS 8, MAXIMUM 2**MAX2
C
C         THE RESULT IS PLACED IN X AS DESCRIBED IN THE
C         INTRODUCTORY TEXT
C
      REAL X(M), BCOS, BSIN, SAVE1, AN, BN, CN, DN, UN, VN,
     $ XN, YN, Z, ZERO, QUART, HALF, ONE, ONE5, TWO, FOUR,
     $ ZATAN, ZFLOAT, ZSIN
C
      DATA MAX2 /21/
      DATA ZERO, QUART, HALF, ONE, ONE5, TWO, FOUR
     $     /0.0, 0.25,  0.5, 1.0,  1.5, 2.0,  4.0/
C
      ZATAN(Z) = ATAN(Z)
      ZFLOAT(K) = FLOAT(K)
      ZSIN(Z) = SIN(Z)
C
C         CHECK FOR VALID TRANSFORM SIZE
C
      II = 8
      DO 2 IPOW = 3, MAX2
      IF (II - M) 1, 4, 3
    1 II = II * 2
    2 CONTINUE
C
C         IF THIS POINT IS REACHED A SIZE ERROR HAS OCCURRED
C
    3 IFAULT = 1
      RETURN
C
C         SEPARATE ODD AND EVEN PARTS INTO TWO HALVES.
C         FIRST BIT-REVERSE THE WHOLE ARRAY OF LENGTH M
C
    4 IFAULT = 0
      CALL SCRAG(X, M, IPOW)
C
C         NEXT BIT-REVERSE THE HALF ARRAYS SEPARATELY
C
      N = M / 2
      JPOW = IPOW - 1
      CALL SCRAG(X, N, JPOW)
      CALL SCRAG(X(N + 1), N, JPOW)
C
C         FASTER ALTERNATIVE REQUIRES AN EXTRA SUBROUTINE.
C         REPLACE TWO PREVIOUS CALLS TO SCRAG WITH ONE TO SCRAM
C             CALL SCRAM(X, X(N + 1), N, JPOW)
C
C         NOW DO THE TRANSFORM
C
      CALL FASTG(X, X(N + 1), N, 1)
C
C         UNSCRAMBLE THE TRANSFORM RESULTS
C
      CALL SCRAG(X, N, JPOW)
      CALL SCRAG(X(N + 1), N, JPOW)
C
C         FASTER ALTERNATIVE REQUIRES AN EXTRA SUBROUTINE.
C         REPLACE TWO PREVIOUS CALLS TO SCRAG WITH ONE TO SCRAM
C             CALL SCRAM(X, X(N + 1), N, JPOW)
```

```
C
      NN = N / 2
C
C        NOW UNRAVEL THE RESULT, FIRST THE SPECIAL CASES
C
      Z = HALF * (X(1) + X(N + 1))
      X(N + 1) = HALF * (X(1) - X(N + 1))
      X(1) = Z
      NN1 = NN + 1
      NN2 = NN1 + N
      X(NN1) = HALF * X(NN1)
      X(NN2) = -HALF * X(NN2)
      Z = FOUR * ZATAN(ONE) / ZFLOAT(N)
      BCOS = -TWO * (ZSIN(Z * HALF) ** 2)
      BSIN = ZSIN(Z)
      UN = ONE
      VN = ZERO
      DO 5 K = 2, NN
      Z = UN * BCOS + VN * BSIN + UN
      VN = VN * BCOS - UN * BSIN + VN
      SAVE1 = ONE5 - HALF * (Z * Z + VN * VN)
      UN = Z * SAVE1
      VN = VN * SAVE1
      KI = N + K
      L = N + 2 - K
      LI = N + L
      AN = QUART * (X(K) + X(L))
      BN = QUART * (X(KI) - X(LI))
      CN = QUART * (X(KI) + X(LI))
      DN = QUART * (X(L) - X(K))
      XN = UN * CN - VN * DN
      YN = UN * DN + VN * CN
      X(K) = AN + XN
      X(KI) = BN + YN
      X(L) = AN - XN
      X(LI) = YN - BN
    5 CONTINUE
      RETURN
      END
C
      SUBROUTINE REVRT(X, M, IFAULT)
C
C        ALGORITHM AS 97.2   APPL. STATIST. (1976) VOL.25, P.166
C
C        INVERSE DISCRETE FOURIER TRANSFORM IN ONE DIMENSION TO
C        GIVE REAL RESULT USING COMPLEX TRANSFORM SUBROUTINE FASTG
C
C        X = ARRAY OF FOURIER COMPONENTS ARRANGED AS RESULTING
C        FROM SUBROUTINE FORRT, TYPE REAL, DIMENSION M.
C        M = THE LENGTH OF THE REAL SEQUENCE WHICH WILL BE FOUND.
C        M MUST BE A POWER OF 2 BETWEEN 2**3 AND 2**MAX2
C
      REAL X(M), BCOS, BSIN, SAVE1, AN, BN, CN, DN, PN, QN,
     $ UN, VN, Z, ZERO, HALF, ONE, ONE5, TWO, FOUR, ZATAN,
     $ ZFLOAT, ZSIN
C
      DATA MAX2 /21/
      DATA ZERO, HALF, ONE, ONE5, TWO, FOUR
     $    /0.0,  0.5, 1.0,  1.5, 2.0,  4.0/
C
      ZATAN(Z) = ATAN(Z)
      ZFLOAT(K) = FLOAT(K)
      ZSIN(Z) = SIN(Z)
C
C        CHECK FOR VALID TRANSFORM SIZE
C
      II = 8
      DO 2 IPOW = 3, MAX2
      IF (II - M) 1, 4, 3
```

```
      1 II = II * 2
      2 CONTINUE
C
C         IF THIS POINT IS REACHED A SIZE ERROR HAS OCCURRED
C
      3 IFAULT = 1
        RETURN
      4 IFAULT = 0
        N = M / 2
        NN = N / 2
C
C         UNDO THIS SPECTRUM INTO THAT OF TWO INTERLEAVED SERIES
C         FIRST THE SPECIAL CASES
C
        Z = X(1) + X(N + 1)
        X(N + 1) = X(1) - X(N + 1)
        X(1) = Z
        NN1 = NN + 1
        NN2 = NN1 + N
        X(NN1) = TWO * X(NN1)
        X(NN2) = -TWO * X(NN2)
        Z = FOUR * ZATAN(ONE) / ZFLOAT(N)
        BCOS = -TWO * (ZSIN(Z * HALF) ** 2)
        BSIN = ZSIN(Z)
        UN = ONE
        VN = ZERO
        DO 5 K = 2, NN
        Z = UN * BCOS + VN * BSIN + UN
        VN = VN * BCOS - UN * BSIN + VN
        SAVE1 = ONE5 - HALF * (Z * Z + VN * VN)
        UN = Z * SAVE1
        VN = VN * SAVE1
        KI = N + K
        L = N + 2 - K
        LI = N + L
        AN = X(K) + X(L)
        BN = X(KI) - X(LI)
        PN = X(K) - X(L)
        QN = X(KI) + X(LI)
        CN = UN * PN + VN * QN
        DN = UN * QN - VN * PN
        X(K) = AN - DN
        X(KI) = BN + CN
        X(L) = AN + DN
        X(LI) = CN - BN
      5 CONTINUE
C
C         NOW DO THE INVERSE TRANSFORM
C
        CALL FASTG(X, X(N + 1), N, -1)
C
C         NOW UNDO THE ORDER - THE HALF ARRAYS ARE ALREADY BIT-REVERSED
C         BIT-REVERSE THE WHOLE ARRAY
C
        CALL SCRAG(X, M, IPOW)
        RETURN
        END
```

Algorithm AS 99

FITTING JOHNSON CURVES BY MOMENTS

By I. D. Hill,   R. Hill[†]   and   R. L. Holder
MRC Clinical   *Wolfson Research*   *Department of Statistics,*
Research Centre,   *Laboratories,*   *University of Birmingham,*
Watford Road, Harrow,   *Queen Elizabeth*   *PO Box 363, Birmingham,*
Middlesex, HA1 3UJ, UK   *Medical Centre,*   *B15 2TT, UK*
                        Birmingham, UK

Keywords: Johnson curves; Transformations of normal curve; Method of moments; Curve fitting.

LANGUAGE

Fortran 66 and 77.

DESCRIPTION AND PURPOSE

Johnson (1949) described a system of frequency curves consisting of:

(1) the lognormal system (or S_L): $z = \gamma + \delta \ln (x - \xi)$, $\xi < x$,
(2) the unbounded system (or S_U): $z = \gamma + \delta \sinh^{-1} ((x - \xi)/\lambda)$,
(3) the bounded system (or S_B): $z = \gamma + \delta \ln ((x - \xi)/(\xi + \lambda - x))$, $\xi < x < \xi + \lambda$,

where z is a standardized normal variable in each case.

For the sake of completeness we have included (4) the normal curve itself; (5) the special case of the S_B curves on the $\beta_2 = \beta_1 + 1$ boundary, which we have called S_T (T standing for 'two-ordinate').

To make the first four moments of x match those of any required distribution it is necessary to determine which of the transformations is required and to evaluate the parameters, γ, δ, λ and ξ.

Fitting by moments is not always a desirable procedure. However, in a number of situations it is quite adequate, without any pretence that it can be regarded as giving the 'best' solution in any sense. In particular, it may be worth while to produce starting values from which to seek for a maximum likelihood solution. Also, moments can sometimes be calculated theoretically, and thus not be subject to sampling error, in which case the objections to fitting by moments do not apply. For discussion of some of the alternative methods of estimating parameters, see Ord (1972).

This algorithm supplements Tables 34, 35 and 36 of Pearson and Hartley (1972), for S_U and S_B curves. These tables are perfectly adequate for many purposes, but interpolation or extrapolation may be hazardous when the required curve is near one of the boundaries.

[†]*Present address:* Data Technology Ltd, PO Box 4T, St Thomas, Barbados, West Indies.

NUMERICAL METHOD

Defining, as is customary, $\sqrt{\beta_1}$, as μ_3/σ^3 and β_2 as μ_4/σ^4, the S_L curves lie on a line in the $\beta_1\beta_2$ plane — thus for these curves β_1 determines β_2. Using ω to denote $\exp(\delta^{-2})$, the S_L β_2 value is found by solving

$$(\omega - 1)(\omega + 2)^2 = \beta_1$$

for ω, and then evaluating

$$\beta_2 = \omega^4 + 2\omega^3 + 3\omega^2 - 3.$$

If the required β_2 is less than this value, S_B (or S_T) is appropriate; if greater, S_U is appropriate.

(1) S_L curves

ω having been evaluated as above,

$$\delta = (\ln \omega)^{-\frac{1}{2}}, \qquad\qquad \gamma = \tfrac{1}{2} \delta \ln \{\omega(\omega - 1)/\mu_2\},$$
$$\xi = \lambda[|\mu_1'| - \exp\{(1/2\delta - \gamma)/\delta\}], \qquad \lambda = \pm 1,$$

where the $\pm$ is determined to be the sign of μ_3. As Johnson (1949) points out, only three parameters are necessary for an S_L curve, but we have found it convenient to include λ as above.

(2) S_U curves

When $\beta_1 = 0$, the required curve is symmetrical, and

$$\omega = \{(2\beta_2 - 2)^{\frac{1}{2}} - 1\}^{\frac{1}{2}}; \quad \delta = (\ln \omega)^{-\frac{1}{2}}; \quad \gamma = 0.$$

For an asymmetrical curve

$$\omega_1 = \{(2\beta_2 - 2\cdot8\beta_1 - 2)^{\frac{1}{2}} - 1\}^{\frac{1}{2}}$$

is taken as a first estimate, and ω, δ and γ found by Johnson's iterative method (Elderton and Johnson, 1969, p. 127). The sign of γ is set to be the opposite of that of μ_3.

In either case ξ and λ are then found from

$$\mu_2 = \tfrac{1}{2} \lambda^2 (\omega - 1)\{\omega \cosh (2\gamma/\delta) + 1\}; \quad \mu_1' = \xi - \lambda\omega^{\frac{1}{2}} \sinh (\gamma/\delta).$$

(3) S_B curves

Approaching the S_T boundary, $\delta \to 0$; approaching the S_L boundary δ tends to the same value as for an S_L curve. A first approximation to δ can be found by interpolating between these two values. The interpolation is made by assuming the shape of the function to be the same at the required β_1 value as it is between the same two δ values when $\beta_1 = 0$. This is well approximated by

$$\delta = (0\cdot626\beta_2 - 0\cdot408)/(3\cdot0 - \beta_2)^{0\cdot479} \text{ if } \beta_2 \geqslant 1\cdot8,$$

and by

$$\delta = 0\cdot8(\beta_2 - 1) \text{ otherwise.}$$

For a given β_1 and first approximation to δ, a first approximation to γ is found using formulae due to Draper (1951).

Evaluation of the first six moments at the given δ and γ values, using Draper's (1952) form of Goodwin's (1949) integral, then enables a two-dimensional Newton—Raphson process to converge on the required values.

Since the first six moments are evaluated at each stage, when the required δ and γ have been found, the first two moments are available to determine λ and ξ.

(4) Normal curves

δ is set to the required value of $1/\sigma$, and γ to $\bar{x}/\sigma$; ξ and λ are set, arbitrarily, to 0.

(5) S_T curves

Since it is unnecessarily complicated to regard these as transformations of the normal curve, totally different meanings of the parameters are used. ξ and λ are set to the two values at which ordinates occur, and δ to the proportion of values at λ. γ is set, arbitrarily, to 0.

STRUCTURE

SUBROUTINE JNSN (XBAR, SD, RB1, BB2, ITYPE, GAMMA, DELTA, XLAM, XI, IFAULT)

Formal parameters

XBAR	Real	input:	the required mean.
SD	Real	input:	the required standard deviation.
RB1	Real	input:	the required value of $\sqrt{\beta_1}$, taking the same sign as the third moment about the mean.
BB2	Real	input:	the required value of β_2; or a negative value to indicate that an S_L curve is desired (or a normal if $\beta_1 = 0$), with the given values of the other three input parameters.
ITYPE	Integer	output:	the type of curve fitted: $1 = S_L, 2 = S_U, 3 = S_B,$ $4 =$ normal, $5 = S_T.$
GAMMA	Real	output:	fitted value of γ.
DELTA	Real	output:	fitted value of δ.
XLAM	Real	output:	fitted value of λ
XI	Real	output:	fitted value of ξ.
IFAULT	Integer	output:	see *failure indications* below.

Failure indications

$IFAULT = 0$ indicates successful completion.
$IFAULT = 1$ indicates a required standard deviation of less than zero.
$IFAULT = 2$ indicates $\beta_2 < \beta_1 + 1$.
$IFAULT = 3$ S_B fitting has failed to converge, so an S_L fit or an S_T fit has been
              made instead. The user should check whether the substituted fit is
              good enough for the purpose.

AUXILIARY ALGORITHMS

Subroutines *SUFIT,* to fit an S_U distribution, *SBFIT,* to fit an S_B distribution,
and *MOM,* to find the first six moments of an S_B distribution are included as
Algorithms AS 99.1, AS 99.2 and AS 99.3 respectively.

MOM may, if desired, be replaced by any standard quadrature routine to
find the first six moments.

PRECISION

Single precision arithmetic is generally sufficient, even on machines that use
only 32 bits for real number representation.

However, if an S_B fit is required close to either the S_L or the S_T boundary,
convergence may not be achieved; double precision working may then be helpful,
if the approximation of taking a distribution on the boundary is regarded as
inadequate. Increasing the values of *LIMIT,* set in *DATA* statements in *SBFIT*
and *MOM* may also be worth considering, provided that the usage of computer
time is not critical.

To produce a double precision version: (i) change the word *REAL* to
DOUBLE PRECISION in each of the four subroutines; (ii) give the real constants
included in *DATA* statements double precision values; (iii) change *ABS, EXP,
ALOG, SIGN* and *SQRT* to *DABS, DEXP, DLOG, DSIGN* and *DSQRT* respectively in the statement functions.

ACCURACY

The parameters found are such that the values of $\sqrt{\beta_1}$ and β_2 achieved are both
within $\pm$ *TOL* of the required values. The value of *TOL* is set in *DATA* statements
in *JNSN, SUFIT* and *SBFIT.* It may be changed if desired but should be identical
in the three places. *TT* in *SBFIT* should be *TOL*^{2}.

The constants *ZZ* and *VV,* set in a *DATA* statement in *MOM,* determine the
accuracy of convergence of the outer and inner loops of the evaluation. *VV*
should be considerably smaller than *ZZ,* which in turn should be smaller than
TT in *SBFIT.*

TIME

On a PDP-11/40, in single precision, typical times to fit an S_L, an S_U, an easy S_B
(midway between the S_L and S_T boundaries) and a difficult S_B (close to a
boundary) are $0 \cdot 028$ sec, $0 \cdot 049$ sec, $2 \cdot 1$ sec and 25 sec respectively.

ADDITIONAL COMMENT

Where one end of the distribution is bounded (S_L), or both ends are bounded (S_B), there may often be a physical reason to know the value(s) of the bound(s). Fitting should then, usually, be performed conditional on the known values of such bounds. The current algorithm does not deal with these cases, but attention is drawn to Bacon-Shone (1985).

EXAMPLE

As an illustration of how algorithms AS 99, AS 100.2 and AS 66 may be put together to estimate tail areas of a distribution whose moments are known, we present the following fragment of program, that will evaluate an approximation to the area above the value C of a χ^2 distribution having F degrees of freedom. We deliberately choose something for which the precise answers are known as a test on the accuracy of the method.

```
C
C        SNV IS ALGORITHM AS 100.2
C        ALNORM IS ALGORITHM AS 66
C
        CALL JNSN(F, SQRT(F + F), SQRT(8.0 / F),
     $   12.0 / F + 3.0, IT, G, D, XL, XI, IFAULT)
        IF (IFAULT .NE. 0) GOTO 100
        A = ALNORM(SNV(C, IT, G, D, XL, XI, IFAULT), .TRUE.)
```

Taking $F = 1, 2, 3, 4$ and values of C for known percentage points, the following results were found:

	Correct result	0·50	0·10	0·01
$F = 1$		0·539	0·0952	0·0105
$F = 2$	Value of	0·512	0·0972	0·0105
$F = 3$	A	0·505	0·0984	0·0104
$F = 4$		0·502	0·0990	0·0104

ACKNOWLEDGEMENTS

We are indebted to J. Draper for permission to make use of his M.Sc. thesis, to the referee for many useful comments, and to R. W. Farebrother for suggesting an improvement to the coding.

EDITORS' REMARKS

The correction noted in AS R33 (Hill and Wheeler, 1981) and a small change to the coding suggested by R. W. Farebrother have been incorporated. The only other changes made to this routine are those facilitating the change to double precision outlined above.

REFERENCES

Bacon-Shone, J. (1985) Algorithm AS 210. Fitting five parameter Johnson S_B curves by moments. *Appl. Statist.*, **34**, in press.

Draper, J. (1951) *Properties of distributions resulting from certain simple transformations of the normal distribution.* M.Sc. thesis. University of London.

Draper, J. (1952) Properties of distributions resulting from certain simple transformations of the normal distribution. *Biometrika*, **39**, 290–301.

Elderton, W. P. and Johnson, N. L. (1969) *Systems of Frequency Curves.* Cambridge: Cambridge University Press.

Goodwin, E. T. (1949) The evaluation of integrals of the form $\int_{-\infty}^{\infty} f(x)e^{-x^2}\,dx$. *Proc. Camb. Phil. Soc.*, **45**, 241–245.

Hill, I. D. and Wheeler, R. E. (1981) Remark AS R33. *Appl. Statist.*, **30**, 106.

Johnson, N. L. (1949) Systems of frequency curves generated by methods of translation. *Biometrika*, **36**, 149–176.

Ord, J. K. (1972) *Families of Frequency Distributions.* London: Griffin.

Pearson, E. S. and Hartley, H. O. (1972) *Biometrika Tables for Statisticians*, Vol. 2. Cambridge: Cambridge University Press.

```
      SUBROUTINE JNSN(XBAR, SD, RB1, BB2, ITYPE, GAMMA, DELTA,
     $ XLAM, XI, IFAULT)
C
C        ALGORITHM AS 99   APPL. STATIST. (1976) VOL.25, P.180
C
C        FINDS TYPE AND PARAMETERS OF A JOHNSON CURVE
C        WITH GIVEN FIRST FOUR MOMENTS
C
      REAL XBAR, SD, RB1, BB2, GAMMA, DELTA, XLAM, XI, TOL,
     $ B1, B2, Y, X, U, W, ZERO, ONE, TWO, THREE, FOUR, HALF,
     $ QUART, ZABS, ZEXP, ZLOG, ZSIGN, ZSQRT
      LOGICAL FAULT
C
      DATA TOL /0.01/
      DATA ZERO, QUART, HALF, ONE, TWO, THREE, FOUR
     $     /0.0,  0.25,  0.5, 1.0, 2.0,   3.0,  4.0/
C
      ZABS(X) = ABS(X)
      ZEXP(X) = EXP(X)
      ZLOG(X) = ALOG(X)
      ZSIGN(X, Y) = SIGN(X, Y)
      ZSQRT(X) = SQRT(X)
C
      IFAULT = 1
      IF (SD .LT. ZERO) RETURN
      IFAULT = 0
      XI = ZERO
      XLAM = ZERO
      GAMMA = ZERO
      DELTA = ZERO
      IF (SD .GT. ZERO) GOTO 10
      ITYPE = 5
      XI = XBAR
      RETURN
   10 B1 = RB1 * RB1
      B2 = BB2
      FAULT = .FALSE.
```

```
C
C         TEST WHETHER LOGNORMAL (OR NORMAL) REQUESTED
C
      IF (B2 .GE. ZERO) GOTO 30
   20 IF (ZABS(RB1) .LE. TOL) GOTO 70
      GOTO 80
C
C         TEST FOR POSITION RELATIVE TO BOUNDARY LINE
C
   30 IF (B2 .GT. B1 + TOL + ONE) GOTO 60
      IF (B2 .LT. B1 + ONE) GOTO 50
C
C         ST DISTRIBUTION
C
   40 ITYPE = 5
      Y = HALF + HALF * ZSQRT(ONE - FOUR / (B1 + FOUR))
      IF (RB1 .GT. ZERO) Y = ONE - Y
      X = SD / ZSQRT(Y * (ONE - Y))
      XI = XBAR - Y * X
      XLAM = XI + X
      DELTA = Y
      RETURN
   50 IFAULT = 2
      RETURN
   60 IF (ZABS(RB1) .GT. TOL .OR. ZABS(B2 - THREE) .GT. TOL) GOTO 80
C
C         NORMAL DISTRIBUTION
C
   70 ITYPE = 4
      DELTA = ONE / SD
      GAMMA = -XBAR / SD
      RETURN
C
C         TEST FOR POSITION RELATIVE TO LOGNORMAL LINE
C
   80 X = HALF * B1 + ONE
      Y = ZABS(RB1) * ZSQRT(QUART * B1 + ONE)
      U = (X + Y) ** (ONE / THREE)
      W = U + ONE / U - ONE
      U = W * W * (THREE + W * (TWO + W)) - THREE
      IF (B2 .LT. ZERO .OR. FAULT) B2 = U
      X = U - B2
      IF (ZABS(X) .GT. TOL) GOTO 90
C
C         LOGNORMAL (SL) DISTRIBUTION
C
      ITYPE = 1
      XLAM = ZSIGN(ONE, RB1)
      U = XLAM * XBAR
      X = ONE / ZSQRT(ZLOG(W))
      DELTA = X
      Y = HALF * X * ZLOG(W * (W - ONE) / (SD * SD))
      GAMMA = Y
      XI = XLAM * (U - ZEXP((HALF / X - Y) / X))
      RETURN
C
C         SB OR SU DISTRIBUTION
C
   90 IF (X .GT. ZERO) GOTO 100
      ITYPE = 2
      CALL SUFIT(XBAR, SD, RB1, B2, GAMMA, DELTA, XLAM, XI)
      RETURN
  100 ITYPE = 3
      CALL SBFIT(XBAR, SD, RB1, B2, GAMMA, DELTA, XLAM, XI, FAULT)
      IF (.NOT. FAULT) RETURN
C
C         FAILURE - TRY TO FIT APPROXIMATE RESULT
C
```

```
      IFAULT = 3
      IF (B2 .GT. B1 + TWO) GOTO 20
      GOTO 40
      END
C
      SUBROUTINE SUFIT (XBAR, SD, RB1, B2, GAMMA, DELTA, XLAM, XI)
C
C        ALGORITHM AS 99.1  APPL. STATIST. (1976) VOL.25, P.180
C
C        FINDS PARAMETERS OF JOHNSON SU CURVE WITH
C        GIVEN FIRST FOUR MOMENTS
C
      REAL XBAR, SD, RB1, B2, GAMMA, DELTA, XLAM, XI, TOL, B1,
     $  B3, W, Y, W1, WM1, Z, V, A, B, X, ZERO, ONE, TWO, THREE,
     $  FOUR, SIX, SEVEN, EIGHT, NINE, TEN, SIXTEN, ZABS, ZEXP,
     $  SIXTEN, ZABS, ZEXP, ZLOG, ZSIGN, ZSQRT
C
      DATA TOL /0.01/
      DATA ZERO,  ONE,   TWO,   THREE, FOUR,  SIX, SEVEN,
     $    EIGHT, NINE,  TEN, SIXTEN, HALF, ONE5,  TWO8,
     $     /0.0,  1.0,   2.0,    3.0,  4.0,  6.0,  7.0,
     $      8.0,  9.0, 10.0,   16.0,  0.5,  1.5,   2.8/
C
      ZABS(X) = ABS(X)
      ZEXP(X) = EXP(X)
      ZLOG(X) = ALOG(X)
      ZSIGN(X, Y) = SIGN(X, Y)
      ZSQRT(X) = SQRT(X)
C
      B1 = RB1 * RB1
      B3 = B2 - THREE
C
C        W IS FIRST ESTIMATE OF EXP(DELTA ** (-2))
C
      W = ZSQRT(ZSQRT(TWO * B2 - TWO8 * B1 - TWO) - ONE)
      IF (ZABS(RB1) .GT. TOL) GOTO 10
C
C        SYMMETRICAL CASE - RESULTS ARE KNOWN
C
      Y = ZERO
      GOTO 20
C
C        JOHNSON ITERATION (USING Y FOR HIS M)
C
   10 W1 = W + ONE
      WM1 = W - ONE
      Z = W1 * B3
      V = W * (SIX + W * (THREE + W))
      A = EIGHT * (WM1 * (THREE + W * (SEVEN + V)) - Z)
      B = SIXTEN * (WM1 * (SIX + V) - B3)
      Y = (ZSQRT(A * A - TWO * B * (WM1 * (THREE + W *
     $  (NINE + W * (TEN + V))) - TWO * W1 * Z)) - A) / B
      Z = Y * WM1 * (FOUR * (W + TWO) * Y + THREE * W1 * W1) ** 2 /
     $  (TWO * (TWO * Y + W1) ** 3)
      V = W * W
      W = ZSQRT(ZSQRT(ONE - TWO * (ONE5 - B2 + (B1 *
     $  (B2 - ONE5 - V * (ONE + HALF * V))) / Z)) - ONE)
      IF (ZABS(B1 - Z) .GT. TOL) GOTO 10
C
C        END OF ITERATION
C
      Y = Y / W
      Y = ZLOG(ZSQRT(Y) + ZSQRT(Y + ONE))
      IF (RB1 .GT. ZERO) Y = -Y
   20 X = ZSQRT(ONE / ZLOG(W))
      DELTA = X
      GAMMA = Y * X
      Y = ZEXP(Y)
      Z = Y * Y
```

```
      X = SD / ZSQRT(HALF * (W - ONE) * (HALF * W *
     $ (Z + ONE / Z) + ONE))
      XLAM = X
      XI = (HALF * ZSQRT(W) * (Y - ONE / Y)) * X + XBAR
      RETURN
      END
C
      SUBROUTINE SBFIT(XBAR, SIGMA, RTB1, B2, GAMMA, DELTA, XLAM,
     $ XI, FAULT)
C
C        ALGORITHM AS 99.2  APPL. STATIST. (1976) VOL.25, P.180
C
C        FINDS PARAMETERS OF JOHNSON SB CURVE WITH
C        GIVEN FIRST FOUR MOMENTS
C
      REAL HMU(6), DERIV(4), DD(4), XBAR, SIGMA, RTB1, B2, GAMMA,
     $ DELTA, XLAM, XI, TT, TOL, RB1, B1, E, U, X, Y, W, F, D,
     $ G, S, H2, T, H2A, H2B, H3, H4, RBET, BET2, ZERO, ONE,
     $ TWO, THREE, FOUR, SIX, HALF, QUART, ONE5, A1, A2, A3,
     $ A4, A5, A6, A7, A8, A9, A10, A11, A12, A13, A14, A15,
     $ A16, A17, A18, A19, A20, A21, A22, ZABS, ZLOG, ZSQRT
      LOGICAL NEG, FAULT
C
      DATA TT, TOL, LIMIT /1.0E-4, 0.01, 50/
      DATA ZERO, ONE, TWO, THREE, FOUR, SIX, HALF, QUART, ONE5
     $      /0.0, 1.0, 2.0,  3.0, 4.0, 6.0, 0.5, 0.25, 1.5/
      DATA    A1,     A2,     A3,     A4,     A5,     A6,
     $        A7,     A8,     A9,     A10,    A11,    A12,
     $        A13,    A14,    A15,    A16,    A17,    A18,
     $        A19,    A20,    A21,    A22
     $   /0.0124, 0.0623, 0.4043,  0.408,  0.479,  0.485,
     $    0.5291, 0.5955,  0.626,   0.64, 0.7077, 0.7466,
     $       0.8, 0.9281, 1.0614,   1.25, 1.7973,    1.8,
     $     2.163,    2.5, 8.5245, 11.346/
C
      ZABS(X) = ABS(X)
      ZLOG(X) = ALOG(X)
      ZSQRT(X) = SQRT(X)
C
      RB1 = ZABS(RTB1)
      B1 = RB1 * RB1
      NEG = RTB1 .LT. ZERO
C
C        GET D AS FIRST ESTIMATE OF DELTA
C
      E = B1 + ONE
      X = HALF * B1 + ONE
      Y = ZABS(RB1) * ZSQRT(QUART * B1 + ONE)
      U = (X + Y) ** (ONE / THREE)
      W = U + ONE / U - ONE
      F = W * W * (THREE + W * (TWO + W)) - THREE
      E = (B2 - E) / (F - E)
      IF (ZABS(RB1) .GT. TOL) GOTO 5
      F = TWO
      GOTO 20
    5 D = ONE / ZSQRT(ZLOG(W))
      IF (D .LT. A10) GOTO 10
      F = TWO - A21 / (D * (D * (D - A19) + A22))
      GOTO 20
   10 F = A16 * D
   20 F = E * F + ONE
      IF (F .LT. A18) GOTO 25
      D = (A9 * F - A4) * (THREE - F) ** (-A5)
      GOTO 30
   25 D = A13 * (F - ONE)
C
C        GET G AS FIRST ESTIMATE OF GAMMA
C
   30 G = ZERO
```

```
      IF (B1 .LT. TT) GOTO 70
      IF (D .GT. ONE) GOTO 40
      G = (A12 * D ** A17 + A8) * B1 ** A6
      GOTO 70
   40 IF (D .LE. A20) GOTO 50
      U = A1
      Y = A7
      GOTO 60
   50 U = A2
      Y = A3
   60 G = B1 ** (U * D + Y) * (A14 + D * (A15 * D - A11))
   70 M = 0
C
C         MAIN ITERATION STARTS HERE
C
   80 M = M + 1
      FAULT = M .GT. LIMIT
      IF (FAULT) RETURN
C
C         GET FIRST SIX MOMENTS FOR LATEST G AND D VALUES
C
      CALL MOM(G, D, HMU, FAULT)
      IF (FAULT) RETURN
      S = HMU(1) * HMU(1)
      H2 = HMU(2) - S
      FAULT = H2 .LE. ZERO
      IF (FAULT) RETURN
      T = ZSQRT(H2)
      H2A = T * H2
      H2B = H2 * H2
      H3 = HMU(3) - HMU(1) * (THREE * HMU(2) - TWO * S)
      RBET = H3 / H2A
      H4 = HMU(4) - HMU(1) * (FOUR * HMU(3) - HMU(1) *
     $   (SIX * HMU(2) - THREE * S))
      BET2 = H4 / H2B
      W = G * D
      U = D * D
C
C         GET DERIVATIVES
C
      DO 120 J = 1, 2
      DO 110 K = 1, 4
      T = K
      IF (J .EQ. 1) GOTO 90
      S = ((W - T) * (HMU(K) - HMU(K + 1)) + (T + ONE) *
     $   (HMU(K + 1) - HMU(K + 2))) / U
      GOTO 100
   90 S = HMU(K + 1) - HMU(K)
  100 DD(K) = T * S / D
  110 CONTINUE
      T = TWO * HMU(1) * DD(1)
      S = HMU(1) * DD(2)
      Y = DD(2) - T
      DERIV(J) = (DD(3) - THREE * (S + HMU(2) * DD(1) - T * HMU(1))
     $   - ONE5 * H3 * Y / H2) / H2A
      DERIV(J + 2) = (DD(4) - FOUR * (DD(3) * HMU(1) + DD(1) * HMU(3))
     $   + SIX * (HMU(2) * T + HMU(1) * (S - T * HMU(1)))
     $   - TWO * H4 * Y / H2) / H2B
  120 CONTINUE
      T = ONE / (DERIV(1) * DERIV(4) - DERIV(2) * DERIV(3))
      U = (DERIV(4) * (RBET - RB1) - DERIV(2) * (BET2 - B2)) * T
      Y = (DERIV(1) * (BET2 - B2) - DERIV(3) * (RBET - RB1)) * T
C
C         FORM NEW ESTIMATES OF G AND D
C
      G = G - U
      IF (B1 .EQ. ZERO .OR. G .LT. ZERO) G = ZERO
      D = D - Y
      IF (ZABS(U) .GT. TT .OR. ZABS(Y) .GT. TT) GOTO 80
```

```
C
C          END OF ITERATION
C
      DELTA = D
      XLAM = SIGMA / ZSQRT(H2)
      IF (NEG) GOTO 130
      GAMMA = G
      GOTO 140
  130 GAMMA = -G
      HMU(1) = ONE - HMU(1)
  140 XI = XBAR - XLAM * HMU(1)
      RETURN
      END
C
      SUBROUTINE MOM(G, D, A, FAULT)
C
C          ALGORITHM AS 99.3  APPL. STATIST. (1976) VOL.25, P.180
C
C          EVALUATES FIRST SIX MOMENTS OF A JOHNSON
C          SB DISTRIBUTION, USING GOODWIN METHOD
C
      REAL A(6), B(6), C(6), G, D, ZZ, VV, RTTWO, RRTPI, W, E, R,
     $  H, T, U, Y, X, V, F, Z, S, P, Q, AA, AB, EXPA, EXPB,
     $  ZERO, QUART, HALF, P75, ONE, TWO, THREE, ZABS, ZEXP
      LOGICAL L, FAULT
C
      DATA ZZ, VV, LIMIT /1.0E-5, 1.0E-8, 500/
C
C          RTTWO IS SQRT(2.0)
C          RRTPI IS RECIPROCAL OF SQRT(PI)
C          EXPA IS A VALUE SUCH THAT EXP(EXPA) DOES NOT QUITE
C             CAUSE OVERFLOW
C          EXPB IS A VALUE SUCH THAT 1.0 + EXP(-EXPB) MAY BE
C             TAKEN TO BE 1.0
C
      DATA     RTTWO,       RRTPI, EXPA, EXPB
     $  /1.414213562, 0.5641895835, 80.0, 23.7/
      DATA ZERO, QUART, HALF, P75, ONE, TWO, THREE
     $    /0.0, 0.25, 0.5, 0.75, 1.0, 2.0,  3.0/
C
      ZABS(X) = ABS(X)
      ZEXP(X) = EXP(X)
C
      FAULT = .FALSE.
      DO 10 I = 1, 6
   10 C(I) = ZERO
      W = G / D
C
C          TRIAL VALUE OF H
C
      IF (W .GT. EXPA) GOTO 140
      E = ZEXP(W) + ONE
      R = RTTWO / D
      H = P75
      IF (D .LT. THREE) H = QUART * D
      K = 1
      GOTO 40
C
C          START OF OUTER LOOP
C
   20 K = K + 1
      IF (K .GT. LIMIT) GOTO 140
      DO 30 I = 1, 6
   30 C(I) = A(I)
C
C          NO CONVERGENCE YET - TRY SMALLER H
C
      H = HALF * H
   40 T = W
```

```
        U = T
        Y = H * H
        X = TWO * Y
        A(1) = ONE / E
        DO 50 I = 2, 6
     50 A(I) = A(I - 1) / E
        V = Y
        F = R * H
        M = 0
C
C          START OF INNER LOOP
C          TO EVALUATE INFINITE SERIES
C
     60 M = M + 1
        IF (M .GT. LIMIT) GOTO 140
        DO 70 I = 1, 6
     70 B(I) = A(I)
        U = U - F
        Z = ONE
        IF (U .GT. -EXPB) Z = ZEXP(U) + Z
        T = T + F
        L = T .GT. EXPB
        IF (.NOT. L) S = ZEXP(T) + ONE
        P = ZEXP(-V)
        Q = P
        DO 90 I = 1, 6
        AA = A(I)
        P = P / Z
        AB = AA
        AA = AA + P
        IF (AA .EQ. AB) GOTO 100
        IF (L) GOTO 80
        Q = Q / S
        AB = AA
        AA = AA + Q
        L = AA .EQ. AB
     80 A(I) = AA
     90 CONTINUE
    100 Y = Y + X
        V = V + Y
        DO 110 I = 1, 6
        IF (A(I) .EQ. ZERO) GOTO 140
        IF (ZABS((A(I) - B(I)) / A(I)) .GT. VV) GOTO 60
    110 CONTINUE
C
C          END OF INNER LOOP
C
        V = RRTPI * H
        DO 120 I = 1, 6
    120 A(I) = V * A(I)
        DO 130 I = 1, 6
        IF (A(I) .EQ. ZERO) GOTO 140
        IF (ZABS((A(I) - C(I)) / A(I)) .GT. ZZ) GOTO 20
    130 CONTINUE
C
C          END OF OUTER LOOP
C
        RETURN
    140 FAULT =.TRUE.
        RETURN
        END
```

Algorithm AS 100

NORMAL–JOHNSON AND JOHNSON–NORMAL TRANSFORMATIONS

By I. D. Hill

MRC Clinical Research Centre, Watford Road, Harrow, Middlesex, HA1 3UJ, UK

Keywords: Random observations; Johnson distributions; Simulation; Monte Carlo methods; Tail areas.

LANGUAGE

Fortran 66 and 77.

DESCRIPTION AND PURPOSE

These algorithms transform a standardized normal variate into a Johnson variate, and vice versa, for a given type of Johnson curve and given values of its parameters, γ, δ, λ and ξ.

If fitting by the first four moments is regarded as adequate, the required type and parameters may be found by using Algorithm AS 99 (Hill *et al.*, 1976).

Two possible uses are: (i) in simulation work to produce pseudo-random observations from distributions of various shapes. Taking a standardized pseudo-random normal deviate as the first parameter of AS 100.1, the output will be a pseudo-random Johnson variate; (ii) if the moments of a sampling distribution are known on the null hypothesis, and an observation of the statistic is available, Algorithm AS 100.2 may be used to perform an approximate significance test, using the method shown in the example given in the text of Algorithm AS 99.

STRUCTURE

REAL FUNCTION AJV (SNV, ITYPE, GAMMA, DELTA, XLAM, XI, IFAULT)

Formal parameters

SNV	Real	input:	the standardized normal variate to be transformed.
ITYPE	Integer	input:	the type of distribution: $1 = S_L$, $2 = S_U$, $3 = S_B$, $4 = $ Normal, as output from Algorithm AS 99.
GAMMA	Real	input:	the value of γ as output from AS 99.
DELTA	Real	input:	the value of δ as output from AS 99.

XLAM	Real	input:	the value of λ as output from **AS 99**.
XI	Real	input:	the value of ξ as output from **AS 99**.
IFAULT	Integer	output:	1 if *ITYPE* is out of range; 0 otherwise.

REAL FUNCTION SNV (AJV, ITYPE, GAMMA, DELTA, XLAM, XI, IFAULT)

Formal parameters

AJV	Real	input:	the Johnson variate to be transformed.

ITYPE, GAMMA, DELTA, XLAM, XI, IFAULT as for AS 100.1 above.

Constant

The constant *C*, set in a *DATA* statement in AS 100.2, is used to switch between the expression $\sqrt{(W^2 + 1)} + W$ and an approximation to it, $-1/2W$, that is numerically preferable when *W* is large and negative. The value of *C* is chosen as the point at which the approximation becomes accurate to four significant figures.

AUXILIARY ALGORITHMS

If using AS 100.1 for simulation purposes, a pseudo-random normal deviate generator is required, to provide values for the *SNV* parameter, algorithm AS 111 (Beasley and Springer, 1977) is suggested, using the output of algorithm AS 183 (Wichmann and Hill, 1982) as its input.

TIME

On a PDP-11/40 the times merely to execute AS 100.1 or AS 100.2 (ignoring the time needed to generate the input parameters) are less than 0·005 sec. These times can be reduced by taking the appropriate parts of the required algorithm and placing them in line, instead of making a procedure call and then branching according to the value of *ITYPE*. Where speed is of the essence this in-line method should be used.

For this reason the algorithms have been written so that the instructions for each type are complete in themselves. Shorter algorithms could be produced by, for example, including *(SNV − GAMMA)/DELTA* only once instead of four times in AS 100.1, but it would then be more difficult to take the instructions needed for one value of *ITYPE* only.

PRECISION

The routine can be converted to double precision by

(1) changing *REAL* to *DOUBLE PRECISION* on both the *FUNCTION* statements and the declarations of variables;

(2) changing the constants in the *DATA* statements to double precision versions;
(3) changing *ABS, EXP, ALOG, SIGN* and *SQRT* to *DABS, DEXP, DLOG, DSIGN* and *DSQRT* in the statement functions.

ACKNOWLEDGEMENT

I thank the referee for insisting that I should expand this contribution to include Algorithm AS 100.2, as well as Algorithm AS 100.1 which I originally intended to present alone.

EDITORS' REMARKS

Remarks AS R33 (Hill and Wheeler, 1981) and AS R48 (Dodgson and Hill, 1983) have been included. The only other changes made to these routines are those facilitating the change to double precision outlined above. The Auxiliary Algorithms section of the description has been updated to refer to more recently published algorithms.

REFERENCES

Beasley, J. D. and Springer, S. G. (1977) Algorithm AS 111. The percentage points of the normal distribution. *Appl. Statist.,* **26,** 118–121. (See also this book, page 188)
Dodgson, J. H. and Hill, I. D. (1983) Remark AS R48. *Appl. Statist.,* **32,** 345.
Elderton, W. P. and Johnson, N. L. (1969) *Systems of Frequency Curves.* Cambridge: Cambridge University Press.
Hill, I. D. (1973) Algorithm AS 66. The Normal integral. *Appl. Statist.,* **22,** 424–427. (See also this book, page 126)
Hill, I. D., Hill, R. and Holder, R. L. (1976) Algorithm AS 99. Fitting Johnson curves by moments. *Appl. Statist.,* **25,** 180–189. (See also this book, page 171)
Hill, I. D. and Wheeler, R. E. (1981) Remark AS R33. *Appl. Statist.,* **30,** 106.
Johnson, N. L. (1949) Systems of frequency curves generated by methods of translation. *Biometrika,* **36,** 149–176.
Wichmann, B. A. and Hill, I. D. (1982) Algorithm AS 183. An efficient and portable pseudo-random number generator. *Appl. Statist.,* **31,** 188–190. (See also this book, page 238)

```
      REAL FUNCTION AJV(SNV, ITYPE, GAMMA, DELTA, XLAM, XI, IFAULT)
C
C         ALGORITHM AS 100.1  APPL. STATIST. (1976) VOL.25, P.190
C
C         CONVERTS A STANDARD NORMAL VARIATE (SNV) TO A
C         JOHNSON VARIATE (AJV)
C
      REAL SNV, GAMMA, DELTA, XLAM, XI, V, W, ZERO, HALF, ONE,
     $ ZABS, ZEXP, ZSIGN
C
      DATA ZERO, HALF, ONE /0.0, 0.5, 1.0/
C
      ZABS(W) = ABS(W)
      ZEXP(W) = EXP(W)
      ZSIGN(W, V) = SIGN(W, V)
C
      AJV = ZERO
      IFAULT = 1
      IF (ITYPE .LT. 1 .OR. ITYPE .GT. 4) RETURN
      IFAULT = 0
      GOTO (10, 20, 30, 40), ITYPE
C
C         SL DISTRIBUTION
C
   10 AJV = XLAM * ZEXP((XLAM * SNV - GAMMA) / DELTA) + XI
      RETURN
C
C         SU DISTRIBUTION
C
   20 W = ZEXP((SNV - GAMMA) / DELTA)
      W = HALF * (W - ONE / W)
      AJV = XLAM * W + XI
      RETURN
C
C         SB DISTRIBUTION
C
   30 W = (SNV - GAMMA) / DELTA
      V = ZEXP(-ZABS(W))
      V = (ONE - V) / (ONE + V)
      AJV = HALF * XLAM * (ZSIGN(V, W) + ONE) + XI
      RETURN
C
C         NORMAL DISTRIBUTION
C
   40 AJV = (SNV - GAMMA) / DELTA
      RETURN
      END
C
      REAL FUNCTION SNV(AJV, ITYPE, GAMMA, DELTA, XLAM, XI, IFAULT)
C
C         ALGORITHM AS 100.2  APPL. STATIST. (1976) VOL.25, P.190
C
C         CONVERTS A JOHNSON VARIATE (AJV) TO A
C         STANDARD NORMAL VARIATE (SNV)
C
      REAL AJV, GAMMA, DELTA, XLAM, XI, V, W, C, ZERO, HALF, ONE,
     $ ZLOG, ZSQRT
C
      DATA ZERO, HALF, ONE, C /0.0, 0.5, 1.0, -63.0/
C
      ZLOG(W) = ALOG(W)
      ZSQRT(W) = SQRT(W)
C
      SNV = ZERO
      IFAULT = 1
      IF (ITYPE .LT. 1 .OR. ITYPE .GT. 4) RETURN
      IFAULT = 0
      GOTO (10, 20, 30, 40), ITYPE
```

```
C
C          SL DISTRIBUTION
C
   10 W = XLAM * (AJV - XI)
      IF (W .LE. ZERO) GOTO 15
      SNV = XLAM * (ZLOG(W) * DELTA + GAMMA)
      RETURN
   15 IFAULT = 2
      RETURN
C
C          SU DISTRIBUTION
C
   20 W = (AJV - XI) / XLAM
      IF (W .GT. C) GOTO 23
      W = -HALF / W
      GOTO 27
   23 W = ZSQRT(W * W + ONE) + W
   27 SNV = ZLOG(W) * DELTA + GAMMA
      RETURN
C
C          SB DISTRIBUTION
C
   30 W = AJV - XI
      V = XLAM - W
      IF (W .LE. ZERO .OR. V .LE. ZERO) GOTO 35
      SNV = ZLOG(W / V) * DELTA + GAMMA
      RETURN
   35 IFAULT = 2
      RETURN
C
C          NORMAL DISTRIBUTION
C
   40 SNV = DELTA * AJV + GAMMA
      RETURN
      END
```

Algorithm AS 111

THE PERCENTAGE POINTS OF THE NORMAL DISTRIBUTION

By J. D. Beasley[†] and S. G. Springer[‡]

Rothamsted Experimental Station

Keywords: Inverse normal; Normal percentage points.

LANGUAGE

Fortran 66 and 77.

DESCRIPTION AND PURPOSE

Given a value of p, this function routine computes the value of x_p for the standard normal distribution. The symbols p and x_p are respectively the lower tail area and its corresponding percentage point. Thus

$$p = \int_{-\infty}^{x_p} (2\pi)^{-\frac{1}{2}} \exp(-t^2/2) \, dt.$$

NUMERICAL METHOD

The subroutine replaces p by $q = p - 0\cdot5$ and then compares $|q|$ with $0\cdot42$. If $|q| \leqslant 0\cdot42$, x_p is formed by a rational approximation

$$x_p = q \cdot A(q^2)/B(q^2),$$

where A and B are polynomials of degrees 3 and 4 respectively, while if $|q| > 0\cdot42$ an auxiliary variable $r = \sqrt{\{\ln(0\cdot5 - |q|)\}}$ is first formed and x_p is then formed as

$$x_p = \pm C(r)/D(r),$$

where C and D are polynomials of degrees 3 and 2 respectively, the sign taken being that of q. The quantity $(0\cdot5 - |q|)$ is formed as either p or $(1 - p)$ to avoid cancellation if p is small.

STRUCTURE

REAL FUNCTION PPND (P, IFAULT)

Formal parameters

P	Real	input:	value of lower tail area p.
IFAULT	Integer	output:	fault indicator.

[†]*Present address:* J. D. Beasley, Tanton and Company Ltd, 7 St James Road, Harpenden, Herts, AL5 4NX, UK.

[‡]*Present address:* Bank of England, Economic Intelligence Dept., Threadneedle Street, London.

Failure indications

$IFAULT = 1$ if $p \leqslant 0$ or $p \geqslant 1$;
$IFAULT = 0$ otherwise.
If $IFAULT = 1$ the value of *PPND* is set equal to zero.

TIME

The function was evaluated on an ICL System 4-70 for $p = 0 \cdot 0001 \ (0 \cdot 0001) \ 1 \cdot 0$ in both single and double precision using both this algorithm and that of AS 70, *GAUINV* (Odeh and Evans, 1974), with the following results:

Average time in milliseconds for one evaluation

	GAUINV	PPND
Single precision	0·49	0·26
Double precision	0·80	0·36

ACCURACY

In the absence of rounding error, the calculated result is the value $x_{p'}$ corresponding to a datum value p' satisfying

$$|p' - p| < 2^{-31}$$

(in other words, the error of the approximation is equivalent to a perturbation of less than 2^{-31} in the datum value). In the presence of rounding error, we may expect to obtain a value $x_{p''}$ corresponding to a p'' satisfying

$$|p'' - p| < \epsilon,$$

where ϵ is of the order of 20 times the value of the smallest digit in the mantissa. For example, a test of 10,000 pseudo-random datum numbers, uniformly distributed on the range $(0, 1)$, on an ICL System 4-70 (hexadecimal arithmetic with truncation, 24-bit mantissa) produced a largest value for ϵ of $1 \cdot 14 \times 10^{-6}$ ($\cong 19 \times 2^{-24}$).

PRECISION

For a double precision version:

(1) Change *REAL* to *DOUBLE PRECISION* on both the *FUNCTION* statement and the declaration of variables.
(2) Change $E0$ to $D0$ throughout the *DATA* statements.
(3) Change the statement functions for *ZABS*, *ZLOG* and *ZSQRT* to use *DABS*, *DLOG* and *DSQRT* respectively.

RELATED ALGORITHMS

This algorithm is identical in specification to *GAUINV* (AS 70), apart from the triviality that we impose no cut-off at $p = 10^{-20}$ and $p = 1 - 10^{-20}$, but appears to be about twice as fast (the precise amount depending on the machine and compiler) and to have no significant compensating disadvantages. In the absence of rounding error, the bound for $|p' - p|$ for AS 70 is approximately 5.92×10^{-9} (compare $2^{-31} \cong 4.66 \times 10^{-10}$), while the test on rounding errors gave a largest value for ϵ of 1.01×10^{-6} (compare 1.14×10^{-6}). It therefore seems reasonable to use this algorithm in place of AS 70 when working in Fortran or in another high-level language.

When working in machine code, the 'fast' method of Marsaglia *et al.* (1964) comes into prominence, particularly in the generation of pseudo-random normal deviates. In a high-level language this method is less effective on account of the fixing and floating involved.

ADDITIONAL COMMENTS

The form of the error bound should be noted. The function is ill-conditioned in the sense that a small change in p produces a larger change in x_p; the ratio of these changes increases without limit as p approaches its end values. The effect of inevitable round-off in the evaluation of the function is equivalent to the effect of some perturbation in the datum value p. Given the inherent ill-conditioning of the function, it is therefore appropriate to state the accuracy of the result in terms of an equivalent perturbation of the datum. If p is the actual datum value and x_q the actual calculated result, then $(x_q - x_p)$ is given to first order in $(q - p)$ by $(q - p) . \exp(x_q^2/2) . \sqrt{(2\pi)}$.

ACKNOWLEDGEMENTS

Our thanks are due to Miss Christine Shelley for her assistance in programming the calculation of the coefficients in this algorithm.

EDITORS' REMARKS

The only change made to this routine is the introduction of statement functions to facilitate a change to double precision.

REFERENCES

Marsaglia, G., MacLaren, M. D. and Bray, T. A. (1964) A fast procedure for generating normal random variables. *Commun. Ass. Comp. Mach.*, 7, 4–10.
Odeh, R. E. and Evans, J. O. (1974) Algorithm AS 70. The percentage points of the normal distribution. *Appl. Statist.*, 23, 96–97.

```
      REAL FUNCTION PPND(P, IFAULT)
C
C        ALGORITHM AS 111   APPL. STATIST. (1977) VOL.26, P.118
C
C        PRODUCES NORMAL DEVIATE CORRESPONDING TO LOWER TAIL AREA OF P.
C        REAL VERSION FOR EPS = 2 ** (-31)
C        THE HASH SUMS ARE THE SUMS OF THE MODULI OF THE COEFFICIENTS.
C        THEY HAVE NO INHERENT MEANINGS BUT ARE INCLUDED FOR USE IN
C        CHECKING TRANSCRIPTIONS.
C        STANDARD FUNCTIONS ABS, ALOG AND SQRT ARE USED.
C
      REAL ZERO, SPLIT, HALF, ONE, A0, A1, A2, A3, B1, B2, B3, B4,
     $ C0, C1, C2, C3, D1, D2, P, Q, R, ZABS, ZLOG, ZSQRT
C
      DATA ZERO, HALF, ONE, SPLIT /0.0E0, 0.5E0, 1.0E0, 0.42E0/
      DATA A0 /         2.50662 82388 4E0/,
     $     A1 /       -18.61500 06252 9E0/,
     $     A2 /        41.39119 77353 4E0/,
     $     A3 /       -25.44106 04963 7E0/,
     $     B1 /        -8.47351 09309 0E0/,
     $     B2 /        23.08336 74374 3E0/,
     $     B3 /       -21.06224 10182 6E0/,
     $     B4 /         3.13082 90983 3E0/
C
C        HASH SUM AB 143.70383 55807 6
C
      DATA C0 /        -2.78718 93113 8E0/,
     $     C1 /        -2.29796 47913 4E0/,
     $     C2 /         4.85014 12713 5E0/,
     $     C3 /         2.32121 27685 8E0/,
     $     D1 /         3.54388 92476 2E0/,
     $     D2 /         1.63706 78189 7E0/
C
C        HASH SUM CD   17.43746 52092 4
C
      ZABS(P) = ABS(P)
      ZLOG(P) = ALOG(P)
      ZSQRT(P) = SQRT(P)
C
      IFAULT = 0
      Q = P - HALF
      IF (ZABS(Q) .GT. SPLIT) GOTO 1
      R = Q * Q
      PPND = Q * (((A3 * R + A2) * R + A1) * R + A0) /
     $  ((((B4 * R + B3) * R + B2) * R + B1) * R + ONE)
      RETURN
    1 R = P
      IF (Q .GT. ZERO) R = ONE - P
      IF (R .LE. ZERO) GOTO 2
      R = ZSQRT(-ZLOG(R))
      PPND = (((C3 * R + C2) * R + C1) * R + C0) /
     $  ((D2 * R + D1) * R + ONE)
      IF (Q .LT. ZERO) PPND = -PPND
      RETURN
    2 IFAULT = 1
      PPND = ZERO
      RETURN
      END
```

Algorithm AS 136

A K-MEANS CLUSTERING ALGORITHM
By J. A. Hartigan and M. A. Wong
Yale University, New Haven, Connecticut, USA

Keywords: K-means clustering; Transfer algorithm.

LANGUAGE
Fortran 66 and 77.

DESCRIPTION AND PURPOSE

The K-means clustering algorithm is described in detail by Hartigan (1975). An efficient version of the algorithm is presented here.

The aim of the K-means algorithm is to divide M points in N dimensions into K clusters so that the within-cluster sum of squares is minimized. It is not practical to require that the solution has minimal sum of squares against all partitions, except when M, N are small and $K = 2$. We seek instead 'local' optima, solutions such that no movement of a point from one cluster to another will reduce the within-cluster sum of squares.

METHOD

The algorithm requires as input a matrix of M points in N dimensions and a matrix of K initial cluster centres in N dimensions. The number of points in cluster L is denoted by $NC(L)$. $D(I, L)$ is the Euclidean distance between point I and cluster L. The general procedure is to search for a K-partition with locally optimal within-cluster sum of squares by moving points from one cluster to another.

Step 1. For each point $I(I = 1, 2, \ldots, M)$, find its closest and second closest cluster centres, $IC1(I)$ and $IC2(I)$ respectively. Assign point I to cluster $IC1(I)$.

Step 2. Update the cluster centres to be the averages of points contained within them.

Step 3. Initially, all clusters belong to the live set.

Step 4. This is the optimal-transfer (*OPTRA*) stage:

Consider each point $I(I = 1, 2, \ldots, M)$ in turn. If cluster $L(L = 1, 2, \ldots, K)$ is updated in the last quick-transfer (*QTRAN*) stage, then it belongs to the live set throughout this stage. Otherwise, at each step, it is not in the live set if it has not been updated in the last M optimal-transfer steps. Let point I be in cluster $L1$. If $L1$ is in the live-set, do *Step* 4a; otherwise, do *Step* 4b.

Step 4a. Compute the minimum of the quantity, $R2 = [NC(L) * D(I,L)^2] / [NC(L) + 1]$, over all clusters $L(L \neq L1, L = 1, 2, \ldots, K)$. Let $L2$ be the cluster with the smallest $R2$. If this value is greater than or equal to $[NC(L1) * D(I,L1)^2] / [NC(L1) - 1]$, no reallocation is necessary and $L2$ is the new $IC2(I)$. (Note that the value $[NC(L1) * D(I,L1)^2]/[NC(L1) - 1]$ is remembered and will remain the same for point I until cluster $L1$ is updated.) Otherwise, point I is allocated to cluster $L2$ and $L1$ is the new $IC2(I)$. Cluster centres are updated to be the means of points assigned to them if reallocation has taken place. The two clusters that are involved in the transfer of point I at this particular step are now in the live set.

Step 4b. This step is the same as *Step* 4a, except that the minimum $R2$ is computed only over clusters in the live set.

Step 5. Stop if the live set is empty. Otherwise, go to *Step* 6 after one pass through the data set.

Step 6. This is the quick transfer (*QTRAN*) stage:

Consider each point $I(I = 1, 2, \ldots, M)$ in turn. Let $L1 = IC1(I)$ and $L2 = IC2(I)$. It is not necessary to check the point I if both the clusters $L1$ and $L2$ have not changed in the last M steps. Compute the values

$$R1 = [NC(L1) * D(I,L1)^2]/[NC(L1)-1] \text{ and}$$

$$R2 = [NC(L2) * D(I,L2)^2]/[NC(L2) + 1].$$

(As noted earlier, $R1$ is remembered and will remain the same until cluster $L1$ is updated.) If $R1$ is less than $R2$, point I remains in cluster $L1$. Otherwise, switch $IC1(I)$ and $IC2(I)$ and update the centres of clusters $L1$ and $L2$. The two clusters are also noted for their involvement in a transfer at this step.

Step 7. If no transfer took place in the last M steps, go to *Step* 4. Otherwise, go to *Step* 6.

STRUCTURE

SUBROUTINE KMNS (A, M, N, C, K, IC1, IC2, NC, AN1, AN2, NCP, D, ITRAN, LIVE, ITER, WSS, IFAULT)

Formal parameters

A	Real array (M, N)	input:	the data matrix.
M	Integer	input:	the number of points.
N	Integer	input:	the number of dimensions.
C	Real array (K, N)	input:	the matrix of initial cluster centres.
		output:	the matrix of final cluster centres.
K	Integer	input:	the number of clusters.
$IC1$	Integer array (M)	output:	the cluster each point belongs to.
$IC2$	Integer array (M)	workspace:	this array is used to remember the cluster which each point is most likely to be transferred to at each step.

NC	Integer array (K)	output:	the number of points in each cluster.
*AN*1	Real array (K)	workspace:	
*AN*2	Real array (K)	workspace:	
NCP	Integer array (K)	workspace:	
D	Real array (M)	workspace:	
ITRAN	Integer array (K)	workspace:	
LIVE	Integer array (K)	workspace:	
ITER	Integer	input:	the maximum number of iterations allowed.
WSS	Real array (K)	output:	the within-cluster sum of squares of each cluster.
IFAULT	Integer	output:	see Fault Diagnostics below.

Fault diagnostics

$IFAULT = 0$ No fault.

$IFAULT = 1$ At least one cluster is empty after the initial assignment. (A better set of initial cluster centres is called for.)

$IFAULT = 2$ The allowed maximum number of iterations is exceeded.

$IFAULT = 3$ K is less than or equal to 1 or greater than or equal to M.

Auxiliary algorithms

The following auxiliary algorithms are called: *SUBROUTINE OPTRA (A, M, N, C, K, IC1, IC2, NC, AN1, AN2, NCP, D, ITRAN, LIVE, INDEX)* and *SUBROUTINE QTRAN (A, M, N, C, K, IC1, IC2, NC, AN1, AN2, NCP, D, ITRAN, INDEX)* which are included.

RELATED ALGORITHMS

A related algorithm is AS 113 (A transfer algorithm for non-hierarchical classification) given by Banfield and Bassill (1977). This algorithm uses swaps as well as transfers to try to overcome the problem of local optima; that is, for all pairs of points, a test is made whether exchanging the clusters to which the points belong will improve the criterion. It will be substantially more expensive than the present algorithm for large M.

The present algorithm is similar to Algorithm AS 58 (Euclidean cluster analysis) given by Sparks (1973). Both algorithms aim at finding a K-partition of the sample, with within-cluster sum of squares which cannot be reduced by moving points from one cluster to the other. However, the implementation of Algorithm AS 58 does not satisfy this condition. At the stage where each point is examined in turn to see if it should be reassigned to a different cluster, only the closest centre is used to check for possible reallocation of the given point; a cluster centre other than the closest one may have the smallest value of the quantity $\{n_l/(n_l + 1)\}d_l^2$, where n_l is the number of points in cluster l and d_l is the distance from cluster l to the given point. Hence, in general, Algorithm AS 58 does not provide a locally optimal solution.

The two algorithms are tested on various generated data sets. The time consumed on the IBM 370/158 and the within-cluster sum of squares of the resulting K-partitions are given in Table 1. While comparing the entries of the table, note that AS 58 does not give locally optimal solutions and so should be expected to take less time. The *WSS* are different for the two algorithms because they arrive at different partitions of the sets of points. A saving of about 50 per cent in time occurs in *KMNS* due to using 'live' sets and due to using a quick-transfer stage which reduces the number of optimal transfer iterations by a factor of 4. Thus, *KMNS* compared to AS 58 is locally optimal and takes less time, especially when the number of clusters is large.

TIME AND ACCURACY

The time is approximately equal to *CMNKI* where *I* is the number of iterations. For an IBM 370/158, $C = 2 \cdot 1 \times 10^{-5}$ sec. However, different data structures require quite different numbers of iterations; and a careful selection of initial cluster centres will also lead to a considerable saving in time.

Storage requirement: $M(N + 3) + K(N + 7)$.

Table 1

		Time (sec)	*WSS*
1. $M = 1000, N = 10, K = 10$	AS 58	63·86	7056·71
(random spherical normal)	*KMNS*	36·66	7065·59
2. $M = 1000, N = 10, K = 10$	AS 58	43·49	7779·70
(two widely separated random normals)	*KMNS*	19·11	7822·01
3. $M = 1000, N = 10, K = 50$	AS 58	135·71	4543·82
(random spherical normal)	*KMNS*	76·00	4561·48
4. $M = 1000, N = 10, K = 50$	AS 58	95·51	5131·04
(two widely separated random normals)	*KMNS*	57·96	5096·23
5. $M = 50, N = 2, K = 8$	AS 58	0·17	21·03
(two widely separated random normals)	*KMNS*	0·18	21·03

Missing variate values cannot be handled by this algorithm.

The algorithm produces a clustering which is only locally optimal; the within-cluster sum of squares may not be decreased by transferring a point from one cluster to another, but different partitions may have the same or smaller within-cluster sum of squares.

The number of iterations required to attain local optimality is usually less than 10.

ADDITIONAL COMMENTS

One way of obtaining the initial cluster centres is suggested here. The points are first ordered by their distances to the overall mean of the sample. Then, for cluster L ($L = 1, 2, \ldots, K$), the $\{1 + (L - 1) * [M/K]\}$th point is chosen to be its initial cluster centre. In effect, some K sample points are chosen as the initial cluster centres. Using this initialization process, it is guaranteed that no cluster will be empty after the initial assignment in the subroutine. A quick initialization, which is dependent on the input order of the points, takes the first K points as the initial centres.

PRECISION

For a double precision version, change *REAL* to *DOUBLE PRECISION* and give double precision versions of the constants in *DATA* statements, in each of the three subroutines. In AS 136, change *FLOAT* to *DFLOAT* if available (though non-standard in Fortran 66). Otherwise change *FLOAT(J)* to *DBLE (FLOAT(J))*.

ACKNOWLEDGEMENT

This research is supported by National Science Foundation Grant MCS75-08374.

EDITORS' REMARKS

Remark AS R39 has been applied. The only other significant changes are those designed to make easy a change to double precision.

REFERENCES

Banfield, C. F. and Bassill, L. C. (1977) Algorithm AS 113. A transfer algorithm for non-hierarchical classification. *Appl. Statist.*, **26**, 206–210.
England, R. and Benyon, D. (1981) Remark AS R39. *Appl. Statist.*, **30**, 355–356.
Hartigan, J. A. (1975) *Clustering Algorithms.* New York: Wiley.
Sparks, D. N. (1973) Algorithm AS 58. Euclidean cluster analysis. *Appl. Statist.*, **22**, 126–130.

```
      SUBROUTINE KMNS(A, M, N, C, K, IC1, IC2, NC, AN1, AN2, NCP,
     $   D, ITRAN, LIVE, ITER, WSS, IFAULT)
C
C
C        ALGORITHM AS 136  APPL. STATIST. (1979) VOL.28, P.100
C
C        DIVIDE M POINTS IN N-DIMENSIONAL SPACE INTO K CLUSTERS
C        SO THAT THE WITHIN-CLUSTER SUM OF SQUARES IS MINIMIZED.
C
      REAL A(M, N), D(M), C(K, N), AN1(K), AN2(K), WSS(K), DT(2),
     $   BIG, AA, DA, DB, DC, TEMP, ONE, ZERO, ZFLOAT
      INTEGER IC1(M), IC2(M), NC(K), NCP(K), ITRAN(K), LIVE(K)
```

```
C
C         DEFINE BIG TO BE A VERY LARGE POSITIVE NUMBER
C
      DATA BIG, ONE, ZERO /1.0E10, 1.0, 0.0/
C
      ZFLOAT(J) = FLOAT(J)
C
      IFAULT = 3
      IF (K .LE. 1 .OR. K .GE. M) RETURN
C
C         FOR EACH POINT I, FIND ITS TWO CLOSEST CENTRES,
C         IC1(I) AND IC2(I). ASSIGN IT TO IC1(I).
C
      DO 55 I = 1, M
      IC1(I) = 1
      IC2(I) = 2
      DO 10 IL = 1, 2
      DT(IL) = ZERO
      DO 10 J = 1, N
      DA = A(I, J) - C(IL, J)
      DT(IL) = DT(IL) + DA * DA
   10 CONTINUE
      IF (DT(1) .LE. DT(2)) GOTO 20
      IC1(I) = 2
      IC2(I) = 1
      TEMP = DT(1)
      DT(1) = DT(2)
      DT(2) = TEMP
   20 IF (K .EQ. 2) GOTO 55
      DO 50 L = 3, K
      DB = ZERO
      DO 30 J = 1, N
      DC = A(I, J) - C(L, J)
      DB = DB + DC * DC
      IF (DB .GE. DT(2)) GOTO 50
   30 CONTINUE
      IF (DB .LT. DT(1)) GOTO 40
      DT(2) = DB
      IC2(I) = L
      GOTO 50
   40 DT(2) = DT(1)
      IC2(I) = IC1(I)
      DT(1) = DB
      IC1(I) = L
   50 CONTINUE
   55 CONTINUE
C
C         UPDATE CLUSTER CENTRES TO BE THE AVERAGE
C         OF POINTS CONTAINED WITHIN THEM
C
      DO 70 L = 1, K
      NC(L) = 0
      DO 60 J = 1, N
   60 C(L, J) = ZERO
   70 CONTINUE
      DO 90 I = 1, M
      L = IC1(I)
      NC(L) = NC(L) + 1
      DO 80 J = 1, N
   80 C(L, J) = C(L, J) + A(I, J)
   90 CONTINUE
C
C         CHECK TO SEE IF THERE IS ANY EMPTY CLUSTER AT THIS STAGE
C
      IFAULT = 1
      DO 100 L = 1, K
      IF (NC(L) .EQ. 0) RETURN
  100 CONTINUE
      IFAULT = 0
```

```
       DO 120 L = 1, K
       AA = NC(L)
       DO 110 J = 1, N
  110 C(L, J) = C(L, J) / AA
C
C          INITIALIZE AN1, AN2, ITRAN AND NCP
C          AN1(L) IS EQUAL TO NC(L) / (NC(L) - 1)
C          AN2(L) IS EQUAL TO NC(L) / (NC(L) + 1)
C          ITRAN(L)=1 IF CLUSTER L IS UPDATED IN THE QUICK-TRANSFER STAGE
C          ITRAN(L)=0 OTHERWISE
C          IN THE OPTIMAL-TRANSFER STAGE, NCP(L) INDICATES THE STEP AT
C          WHICH CLUSTER L IS LAST UPDATED
C          IN THE QUICK-TRANSFER STAGE, NCP(L) IS EQUAL TO THE STEP AT
C          WHICH CLUSTER L IS LAST UPDATED PLUS M
C
       AN2(L) = AA / (AA + ONE)
       AN1(L) = BIG
       IF (AA .GT. ONE) AN1(L) = AA / (AA - ONE)
       ITRAN(L) = 1
       NCP(L) = -1
  120 CONTINUE
       INDEX = 0
       DO 140 IJ = 1, ITER
C
C          IN THIS STAGE, THERE IS ONLY ONE PASS THROUGH THE DATA.
C          EACH POINT IS REALLOCATED, IF NECESSARY, TO THE CLUSTER
C          THAT WILL INDUCE THE MAXIMUM REDUCTION IN WITHIN-CLUSTER
C          SUM OF SQUARES
C
       CALL OPTRA(A, M, N, C, K, IC1, IC2, NC, AN1, AN2, NCP,
     $  D, ITRAN, LIVE, INDEX)
C
C          STOP IF NO TRANSFER TOOK PLACE IN THE LAST M
C          OPTIMAL-TRANSFER STEPS
C
       IF (INDEX .EQ. M) GOTO 150
C
C          EACH POINT IS TESTED IN TURN TO SEE IF IT SHOULD BE
C          REALLOCATED TO THE CLUSTER WHICH IT IS MOST LIKELY TO
C          BE TRANSFERRED TO (IC2(I)) FROM ITS PRESENT CLUSTER (IC1(I)).
C          LOOP THROUGH THE DATA UNTIL NO FURTHER CHANGE IS TO TAKE PLACE
C
       CALL QTRAN(A, M, N, C, K, IC1, IC2, NC, AN1, AN2,
     $  NCP, D, ITRAN, INDEX)
C
C          IF THERE ARE ONLY TWO CLUSTERS,
C          NO NEED TO RE-ENTER OPTIMAL-TRANSFER STAGE
C
       IF (K .EQ. 2) GOTO 150
C
C          NCP HAS TO BE SET TO 0 BEFORE ENTERING OPTRA
C
       DO 130 L = 1, K
  130 NCP(L) = 0
  140 CONTINUE
C
C          SINCE THE SPECIFIED NUMBER OF ITERATIONS IS EXCEEDED
C          IFAULT IS SET TO BE EQUAL TO 2.
C          THIS MAY INDICATE UNFORESEEN LOOPING
C
       IFAULT = 2
C
C          COMPUTE WITHIN-CLUSTER SUM OF SQUARES FOR EACH CLUSTER
C
  150 DO 160 L = 1, K
       WSS(L) = ZERO
       DO 160 J = 1, N
       C(L, J) = ZERO
```

```
  160 CONTINUE
      DO 170 I = 1, M
      II = IC1(I)
      DO 170 J = 1, N
      C(II, J) = C(II, J) + A(I, J)
  170 CONTINUE
      DO 190 J = 1, N
      DO 180 L = 1, K
  180 C(L, J) = C(L, J) / ZFLOAT(NC(L))
      DO 190 I = 1, M
      II = IC1(I)
      DA = A(I, J) - C(II, J)
      WSS(II) = WSS(II) + DA * DA
  190 CONTINUE
      RETURN
      END
C
      SUBROUTINE OPTRA(A, M, N, C, K, IC1, IC2, NC, AN1,
     $ AN2, NCP, D, ITRAN, LIVE, INDEX)
C
C        ALGORITHM AS 136.1  APPL. STATIST. (1979) VOL.28, P.100
C
C        THIS IS THE OPTIMAL-TRANSFER STAGE
C
C        EACH POINT IS REALLOCATED, IF NECESSARY, TO THE
C        CLUSTER THAT WILL INDUCE A MAXIMUM REDUCTION IN
C        THE WITHIN-CLUSTER SUM OF SQUARES
C
      REAL A(M, N), D(M), C(K, N), AN1(K), AN2(K), BIG, DA, DB,
     $ DC, DD, DE, DF, RR, R2, AL1, AL2, ALT, ALW, ONE, ZERO
      INTEGER IC1(M), IC2(M), NC(K), NCP(K), ITRAN(K), LIVE(K)
C
C        DEFINE BIG TO BE A VERY LARGE POSITIVE NUMBER
C
      DATA BIG, ONE, ZERO /1.0E10, 1.0, 0.0/
C
C        IF CLUSTER L WAS UPDATED IN THE LAST QUICK-TRANSFER STAGE,
C        IT BELONGS TO THE LIVE SET THROUGHOUT THIS STAGE.
C        OTHERWISE, AT EACH STEP, IT IS NOT IN THE LIVE SET IF IT
C        HAS NOT BEEN UPDATED IN THE LAST M OPTIMAL-TRANSFER STEPS
C
      DO 10 L = 1, K
      IF (ITRAN(L) .EQ. 1) LIVE(L) = M + 1
   10 CONTINUE
      DO 100 I = 1, M
      INDEX = INDEX + 1
      L1 = IC1(I)
      L2 = IC2(I)
      LL = L2
C
C        IF POINT I IS THE ONLY MEMBER OF CLUSTER L1, NO TRANSFER
C
      IF (NC(L1) .EQ. 1) GOTO 90
C
C        IF L1 HAS NOT YET BEEN UPDATED IN THIS STAGE
C        NO NEED TO RECOMPUTE D(I)
C
      IF (NCP(L1) .EQ. 0) GOTO 30
      DE = ZERO
      DO 20 J = 1, N
      DF = A(I, J) - C(L1, J)
      DE = DE + DF * DF
   20 CONTINUE
      D(I) = DE * AN1(L1)
C
C        FIND THE CLUSTER WITH MINIMUM R2
C
   30 DA = ZERO
```

```
            DO 40 J = 1, N
            DB = A(I, J) - C(L2, J)
            DA = DA + DB * DB
   40 CONTINUE
            R2 = DA * AN2(L2)
            DO 60 L = 1, K
C
C           IF I IS GREATER THAN OR EQUAL TO LIVE(L1), THEN L1 IS
C           NOT IN THE LIVE SET.  IF THIS IS TRUE, WE ONLY NEED TO
C           CONSIDER CLUSTERS THAT ARE IN THE LIVE SET FOR POSSIBLE
C           TRANSFER OF POINT I.  OTHERWISE, WE NEED TO CONSIDER
C           ALL POSSIBLE CLUSTERS
C
            IF (I .GE. LIVE(L1) .AND. I .GE. LIVE(L) .OR.
         $  L .EQ. L1 .OR. L .EQ. LL) GOTO 60
            RR = R2 / AN2(L)
            DC = ZERO
            DO 50 J = 1, N
            DD = A(I, J) - C(L, J)
            DC = DC + DD * DD
            IF (DC .GE. RR) GOTO 60
   50 CONTINUE
            R2 = DC * AN2(L)
            L2 = L
   60 CONTINUE
            IF (R2 .LT. D(I)) GOTO 70
C
C           IF NO TRANSFER IS NECESSARY, L2 IS THE NEW IC2(I)
C
            IC2(I) = L2
            GOTO 90
C
C           UPDATE CLUSTER CENTRES, LIVE, NCP, AN1 AND AN2
C           FOR CLUSTERS L1 AND L2, AND UPDATE IC1(I) AND IC2(I)
C
   70 INDEX = 0
            LIVE(L1) = M + I
            LIVE(L2) = M + I
            NCP(L1) = I
            NCP(L2) = I
            AL1 = NC(L1)
            ALW = AL1 - ONE
            AL2 = NC(L2)
            ALT = AL2 + ONE
            DO 80 J = 1, N
            C(L1, J) = (C(L1, J) * AL1 - A(I, J)) / ALW
            C(L2, J) = (C(L2, J) * AL2 + A(I, J)) / ALT
   80 CONTINUE
            NC(L1) = NC(L1) - 1
            NC(L2) = NC(L2) + 1
            AN2(L1) = ALW / AL1
            AN1(L1) = BIG
            IF (ALW .GT. ONE) AN1(L1) = ALW / (ALW - ONE)
            AN1(L2) = ALT / AL2
            AN2(L2) = ALT / (ALT + ONE)
            IC1(I) = L2
            IC2(I) = L1
   90 IF (INDEX .EQ. M) RETURN
  100 CONTINUE
            DO 110 L = 1, K
C
C           ITRAN(L) IS SET TO ZERO BEFORE ENTERING QTRAN.
C           ALSO, LIVE(L) HAS TO BE DECREASED BY M BEFORE
C           RE-ENTERING OPTRA
C
            ITRAN(L) = 0
            LIVE(L) = LIVE(L) - M
  110 CONTINUE
            RETURN
            END
```

```
C
      SUBROUTINE QTRAN(A, M, N, C, K, IC1, IC2, NC, AN1,
     $ AN2, NCP, D, ITRAN, INDEX)
C
C         ALGORITHM AS 136.2  APPL. STATIST. (1979) VOL.28, P.100
C
C         THIS IS THE QUICK-TRANSFER STAGE.
C         IC1(I) IS THE CLUSTER WHICH POINT I BELONGS TO.
C         IC2(I) IS THE CLUSTER WHICH POINT I IS MOST
C         LIKELY TO BE TRANSFERRED TO.
C         FOR EACH POINT I, IC1(I) AND IC2(I) ARE SWITCHED, IF
C         NECESSARY, TO REDUCE WITHIN-CLUSTER SUM OF SQUARES.
C         THE CLUSTER CENTRES ARE UPDATED AFTER EACH STEP
C
      REAL A(M, N), D(M), C(K, N), AN1(K), AN2(K), BIG, DA, DB,
     $ DD, DE, R2, AL1, AL2, ALT, ALW, ONE, ZERO
      INTEGER IC1(M), IC2(M), NC(K), NCP(K), ITRAN(K)
C
C         DEFINE BIG TO BE A VERY LARGE POSITIVE NUMBER
C
      DATA BIG, ONE, ZERO /1.0E10, 1.0, 0.0/
C
C         IN THE OPTIMAL-TRANSFER STAGE, NCP(L) INDICATES THE
C         STEP AT WHICH CLUSTER L IS LAST UPDATED
C         IN THE QUICK-TRANSFER STAGE, NCP(L) IS EQUAL TO THE
C         STEP AT WHICH CLUSTER L IS LAST UPDATED PLUS M
C
      ICOUN = 0
      ISTEP = 0
   10 DO 70 I = 1, M
      ICOUN = ICOUN + 1
      ISTEP = ISTEP + 1
      L1 = IC1(I)
      L2 = IC2(I)
C
C         IF POINT I IS THE ONLY MEMBER OF CLUSTER L1, NO TRANSFER
C
      IF (NC(L1) .EQ. 1) GOTO 60
C
C         IF ISTEP IS GREATER THAN NCP(L1), NO NEED TO RECOMPUTE
C         DISTANCE FROM POINT I TO CLUSTER L1
C         NOTE THAT IF CLUSTER L1 IS LAST UPDATED EXACTLY M STEPS
C         AGO WE STILL NEED TO COMPUTE THE DISTANCE FROM POINT I
C         TO CLUSTER L1
C
      IF (ISTEP .GT. NCP(L1)) GOTO 30
      DA = ZERO
      DO 20 J = 1, N
      DB = A(I, J) - C(L1, J)
      DA = DA + DB * DB
   20 CONTINUE
      D(I) = DA * AN1(L1)
C
C         IF ISTEP IS GREATER THAN OR EQUAL TO BOTH NCP(L1) AND
C         NCP(L2) THERE WILL BE NO TRANSFER OF POINT I AT THIS STEP
C
   30 IF (ISTEP .GE. NCP(L1) .AND. ISTEP .GE. NCP(L2)) GOTO 60
      R2 = D(I) / AN2(L2)
      DD = ZERO
      DO 40 J = 1, N
      DE = A(I, J) - C(L2, J)
      DD = DD + DE * DE
      IF (DD .GE. R2) GOTO 60
   40 CONTINUE
C
C         UPDATE CLUSTER CENTRES, NCP, NC, ITRAN, AN1 AND AN2
C         FOR CLUSTERS L1 AND L2.  ALSO, UPDATE IC1(I) AND IC2(I).
C         NOTE THAT IF ANY UPDATING OCCURS IN THIS STAGE,
C         INDEX IS SET BACK TO 0
C
```

```
      ICOUN = 0
      INDEX = 0
      ITRAN(L1) = 1
      ITRAN(L2) = 1
      NCP(L1) = ISTEP + M
      NCP(L2) = ISTEP + M
      AL1 = NC(L1)
      ALW = AL1 - ONE
      AL2 = NC(L2)
      ALT = AL2 + ONE
      DO 50 J = 1, N
      C(L1, J) = (C(L1, J) * AL1 - A(I, J)) / ALW
      C(L2, J) = (C(L2, J) * AL2 + A(I, J)) / ALT
   50 CONTINUE
      NC(L1) = NC(L1) - 1
      NC(L2) = NC(L2) + 1
      AN2(L1) = ALW / AL1
      AN1(L1) = BIG
      IF (ALW .GT. ONE) AN1(L1) = ALW / (ALW - ONE)
      AN1(L2) = ALT / AL2
      AN2(L2) = ALT / (ALT + ONE)
      IC1(I) = L2
      IC2(I) = L1
C
C        IF NO REALLOCATION TOOK PLACE IN THE LAST M STEPS, RETURN
C
   60 IF (ICOUN .EQ. M) RETURN
   70 CONTINUE
      GOTO 10
      END
```

Algorithm AS 147

A SIMPLE SERIES FOR THE INCOMPLETE GAMMA INTEGRAL

By Chi-Leung (Andy) Lau

Dept. of Industrial and Management Engineering,
Montana State University,
Bozeman, Montana, USA

Present address: China Project, Americas/Pacific Operations, Tektronix Inc., Beaverton, Oregon, USA.

Keywords: Incomplete gamma integral; Natural logarithm of gamma function.

LANGUAGE

Fortran 66 and 77.

DESCRIPTION AND PURPOSE

Purpose

To evaluate the incomplete gamma integral,

$$G(y, p) = \frac{1}{\Gamma(p)} \int_0^y t^{p-1} e^{-t} \, dt, \quad p, y > 0.$$

The present algorithm is different from algorithm AS 32 in that AS 32 considers separate cases and employs continued fraction approximation, whereas in this algorithm, there is only one case. The programming effort is substantially reduced without sacrificing accuracy.

Theory

Set

$$L(y, p) = \int_0^1 t^{p-1} e^{-yt} \, dt, \quad p, y > 0.$$

$L(y, p)$ then enjoys the following recursive property

$$L(y, p) = \frac{e^{-y}}{p} + \frac{y}{p} L(y, p + 1). \qquad (*)$$

Note that as $p \to \infty$, $L(y, p) \to 0$. Using (*) repeatedly, we get

$$L(y, p) = \frac{e^{-y}}{p} \sum_{n=0}^{\infty} \left\{ \prod_{i=1}^{n} \left(\frac{y}{p+i} \right) \right\} = \frac{e^{-y}}{p} \sum_{n=0}^{\infty} C_n(y, p),$$

where

$$C_0(y, p) = 1 \quad \text{and} \quad C_n(y, p) = \frac{y}{p+n} C_{n-1}(y, p), \quad n = 1, 2, \ldots .$$

Now,

$$G(y, p) = \frac{1}{\Gamma(p)} \int_0^y u^{p-1} e^{-u} \, du = \frac{1}{\Gamma(p)} \int_0^1 (yt)^{p-1} e^{-yt} y \, dt$$

$$= \frac{y^p}{\Gamma(p)} L(y, p) = f \sum_{n=0}^{\infty} C_n(y, p)$$

with

$$f = \frac{y^p}{\Gamma(p+1) e^y} .$$

STRUCTURE

REAL FUNCTION GAMMDS (Y, P, IFAULT)

Formal parameters

Y	Real	input.	the value of upper integral limit y.
P	Real	input:	the value of the parameter p.
IFAULT	Integer	output:	0 no error;
			1 if y or $p \leqslant 0$;
			2 if $f = 0$
			(f defined as above).

AUXILIARY ALGORITHM

The natural logarithm of the gamma function $\Gamma(p)$, p positive, is assumed available, e.g. Hastings's, formula or Stirling's approximation. As presented, the routine uses *ALOGAM* of Pike and Hill (1966).

ACCURACY

The accuracy of the result should normally be not less than the value of E, but due to truncation and rounding errors the last digit may not be correct. Thus to increase the accuracy, reduce the value of E in the *DATA* statement. When the true answer is very close to $1 \cdot 0$, the algorithm will return a result of about $1 \cdot 0 - E$.

PRECISION

The routine can be converted to double precision by changing *REAL FUNCTION* to *DOUBLE PRECISION FUNCTION* and *REAL* to *DOUBLE PRECISION,* replacing the real constants in the *DATA* statement by double precision values, and changing the statement functions to use *DEXP* and *DLOG*.

EDITORS' REMARKS

The only changes made to this routine are those facilitating the change to double precision outlined above.

REFERENCES

Bhattacharjee, G. P. (1970) Algorithm AS 32. The incomplete gamma integral. *Appl. Statist.*, **19**, 285–287.

Hastings, C. J. (1955) *Approximations for Digital Computers.* Princeton, N.J.: Princeton University Press.

Pike, M. C. and Hill, I. D. (1966) Algorithm 291. Logarithm of the gamma function. *Commun. Ass. Comput. Mach.*, **9**, 684. (See also this book, page 243)

```
      REAL FUNCTION GAMMDS(Y, P, IFAULT)
C
C         ALGORITHM AS 147   APPL. STATIST. (1980) VOL.29, P.113
C
C         COMPUTES THE INCOMPLETE GAMMA INTEGRAL FOR POSITIVE
C         PARAMETERS Y, P USING AN INFINITE SERIES.
C
      REAL Y, P, E, ONE, ZERO, A, C, F, ALOGAM, ZEXP, ZLOG
C
      DATA E, ZERO, ONE /1.0E-6, 0.0, 1.0/
C
      ZEXP(A) = EXP(A)
      ZLOG(A) = ALOG(A)
C
C         CHECKS ADMISSIBILITY OF ARGUMENTS AND VALUE OF F
C
      IFAULT = 1
      GAMMDS = ZERO
      IF (Y .LE. ZERO .OR. P .LE. ZERO) RETURN
      IFAULT = 2
C
C         ALOGAM IS NATURAL LOG OF GAMMA FUNCTION
C         NO NEED TO TEST IFAIL AS AN ERROR IS IMPOSSIBLE
C
      F = ZEXP(P * ZLOG(Y) - ALOGAM(P + ONE, IFAIL) - Y)
      IF (F .EQ. ZERO) RETURN
      IFAULT = 0
C
C         SERIES BEGINS
C
      C = ONE
      GAMMDS = ONE
      A = P
    1 A = A + ONE
      C = C * Y / A
      GAMMDS = GAMMDS + C
      IF (C / GAMMDS .GT. E) GOTO 1
      GAMMDS = GAMMDS * F
      RETURN
      END
```

Algorithm AS 154

AN ALGORITHM FOR EXACT MAXIMUM LIKELIHOOD ESTIMATION OF AUTOREGRESSIVE–MOVING AVERAGE MODELS BY MEANS OF KALMAN FILTERING

By

G. Gardner   A. C. Harvey   and   G. D. A. Phillips

Service in   *Department of Statistics,*   *School of Economic Studies,*
Informatics and   *London School of*   *University of Leeds,*
Analysis Ltd,   *Economics,*   *Leeds, LS2 9JT, UK*
London, UK   *Houghton Street,*
                        London, WC2A 2AE, UK

Keywords: Maximum likelihood; Autoregressive–moving average model; Kalman filter.

LANGUAGE

Fortran 66 and 77.

DESCRIPTION AND PURPOSE

The algorithm presented here enables the *exact* likelihood function of a stationary autoregressive–moving average (*ARMA*) process to be calculated by means of the Kalman filter; see Harvey and Phillips (1976, 1979). Two subroutines are basic to the algorithm. The first, subroutine *STARMA*, casts the *ARMA* model into the 'state space' form necessary for Kalman filtering, and computes the covariance matrix associated with the initial value of the state vector. The second subroutine, *KARMA*, carries out the recursions and produces a set of standardized prediction errors, together with the determinant of the covariance matrix of the observations. These two quantities together yield the exact likelihood, and this may be maximized by an iterative procedure based on a numerical optimization algorithm which does not require analytic derivatives.

Subroutine *KARMA* contains a device whereby the likelihood may be approximated to a level of accuracy which is under the control of the user. This enables a considerable amount of computing time to be saved, with very little attendant loss in precision.

Finally, another subroutine, *KALFOR*, may be used to compute predictions of future values of the series, together with the associated conditional mean square errors.

THEORY

An autoregressive–moving average process is defined by

$$w_t = \phi_1 w_{t-1} + \dots \phi_p \, w_{t-p} + \epsilon_t + \theta_1 \epsilon_{t-1} + \dots \theta_q \epsilon_{t-q}, \ t = 1,\dots,n, \quad (1)$$

where the ϵ_t's are normally and independently distributed with mean zero and variance σ^2, and w_t is observable. Such a process will be referred to as an $ARMA(p, q)$ process and the set of parameters $(\phi_1,...,\phi_p, \theta_1,...,\theta_q)$ will be denoted by (ϕ, θ).

An $ARMA(p, q)$ process may be put in 'state space' form by defining an $r \times 1$ vector, α_t, which obeys the 'transition equation'

$$\alpha_t = T\alpha_{t-1} + R\epsilon_t, \quad t = 1,...,n, \tag{2}$$

where $r = \max(p, q + 1)$ and

$$T = \begin{bmatrix} \phi_1 & & \\ \vdots & & I_{r-1} \\ \phi_{r-1} & & \\ \phi_r & & O'_{r-1} \end{bmatrix}, \quad \text{and } R = \begin{bmatrix} 1 \\ \theta_1 \\ \vdots \\ \\ \theta_{r-1} \end{bmatrix}. \tag{3}$$

Note that, unless $p = q + 1$, some of the ϕ_is or θ_js will be identically equal to zero. The associated 'measurement equation' is

$$w_t = (1 \; O'_{r-1})\alpha_t = z'_t \alpha_t, \quad t = 1,...,n. \tag{4}$$

Equations (2) and (4) constitute a linear dynamic model. Given a_{t-1}, an estimate of the state vector at time $t - 1$, together with a matrix P_{t-1} defined by

$$E[(a_{t-1} - \alpha_{t-1})(a_{t-1} - \alpha_{t-1})'] = \sigma^2 P_{t-1},$$

a prediction of α_t, $a_{t|t-1}$, may be made. This may then be updated once the tth observation, w_t, becomes available. The prediction and updating are carried out by means of a set of recursive equations known as the 'Kalman filter'. The parameter σ^2 does not appear in these recursions.

In order to start the recursions, an initial estimator of the state α_0 is needed, together with the associated matrix P_0. The best estimator of α_0 for the $ARMA$ model is $a_0 = 0$, and the matrix P_0 is therefore given by $\sigma^{-2}E[\alpha_0 \; \alpha_0']$. The evaluation of P_0 constitutes a key feature of the present algorithm, and the method employed is discussed at some length in the next section.

Application of the recursive formulae yields a set of n standardized residuals, denoted by $\tilde{\nu}_t$, $t = 1,...,n$, together with a set of n quantities, f_t, $t = 1,...,n$, proportional to the one-step prediction mean square errors. The log-likelihood function may then be maximized with respect to (ϕ, θ) by minimizing

$$L^*(\phi, \theta) = n \log S(\phi, \theta) + \sum_{t=1}^{n} \log f_t, \tag{5}$$

where $S(\phi, \theta) = \Sigma \tilde{\nu}_t^2$. The subroutine $KARMA$ outputs the second term in expression (5) as $SUMLOG$, and $S(\phi, \theta)$ as SSQ.

An approximation to the likelihood may be obtained as follows. Once a certain number of observations, say t^*, have been processed, future values of $\tilde{v}_t$ are approximated by $\hat{v}_t$ which is obtained directly from the $ARMA$ equation (1), i.e.

$$\hat{v}_t = w_t - \phi_1 w_{t-1} - \ldots \phi_p w_{t-p} - \theta_1 \hat{v}_{t-1} - \ldots \theta_q \hat{v}_{t-q}, \ t = t^* + 1, t^* + 2, \ldots, \quad (6)$$

where $\hat{v}_t = \tilde{v}_t, \ t = t^*, \ldots, t^* - q + 1$. The value of t^*, the point at which the switch to the 'quick recursions' takes place, is determined automatically as soon as $f_t < 1 + \delta$. The choice of δ, which will generally be a small positive number, say $0 \cdot 01$ or $0 \cdot 001$, is open to the user. If δ is set equal to a negative number the full Kalman filter is carried out for all observations and the exact likelihood is obtained. Results concerning the trade-off between accuracy and computational efficiency for the approximation are given in Table 1. The figures show the time taken to compute the likelihood function for several values of θ, the parameter in a first-order moving average process. The results indicate that setting δ equal to a value of, say, $0 \cdot 01$ or $0 \cdot 001$ yields a negligible error of approximation while saving a considerable amount of computing time. This saving is particularly marked when $|\theta|$ is relatively small. With $\theta = 0 \cdot 5$, for example, the likelihood function may be computed very accurately in a time which is only marginally greater than that needed to compute the conditional sum of squares.

Table 1

The evaluation of the likelihood for a $MA(1)$ model by the modified Kalman filter method with different values of δ

		$n = 20$			$n = 60$		
		$\theta = 0 \cdot 5$	$\theta = 0 \cdot 8$	$\theta = 0 \cdot 99$	$\theta = 0 \cdot 5$	$\theta = 0 \cdot 8$	$\theta = 0 \cdot 99$
Exact	L^*	66·3606	66·9910	68·8726	255·643	255·903	259·166
likelihood	Time[†]	0·36	0·36	0·35	0·95	0·95	0·94
$\delta = 0 \cdot 001$	L^*	66·3601	66·9906	–	255·641	255·903	–
	t^*	4	13	–	4	13	–
	Time[†]	0·25	0·31	–	0·55	0·62	–
$\delta = 0 \cdot 01$	L^*	66·3541	66·9816	–[‡]	255·636	255·909	259·116
	t^*	3	8	–	3	8	54
	Time[†]	0·24	0·27	–	0·55	0·58	0·91
$\delta = 0 \cdot 1$	L^*	66·2636	66·7719	68·8116	255·704	255·794	259·142
	t^*	1	3	9	1	3	9
	Time[†]	0·23	0·24	0·28	0·53	0·55	0·58
Conditional	L^*	66·0853	65·9915	66·4512	256·528	257·729	264·634
sum of							
squares	Time[†]	0·22	0·22	0·21	0·53	0·52	0·51

[†] Number of seconds on an ICL 4130 computer.

[‡] A '–' indicates that no switch occurred, i.e. $f_t > 1 + \delta$ for all t.

Once estimates of ϕ and θ have been obtained, one may wish to obtain predictions of future values of the series. The predicted value of w_{n+m}, and its mean square error, conditional on (ϕ, θ), are obtained from the recursions

$$a_{n+t|n} = Ta_{n+t-1|n},$$

$$P_{n+t|n} = TP_{n+t-1|n} T' + RR', \quad t = 1,\ldots,m, \tag{7}$$

where $a_{n|n} = a_n$ and $P_{n|n} = P_n$. The first element of $a_{n+m|n}$ is the predicted value of w_{n+m}, while the top left-hand element of $P_{n+m|n}$ gives the associated conditional mean square error when multiplied by an estimate of σ^2.

METHOD

The initial matrix, P_0, obeys the equation

$$P_0 = TP_0 T' + RR'. \tag{8}$$

If $V = RR'$, and if p_{ij}, t_{ij} and v_{ij} denote the element in the ith row and jth column of P_0, T and V respectively then

$$p_{ij} = \sum_k \sum_l t_{ik} p_{kl} t_{jl} + v_{ij}, \tag{9}$$

i.e.

$$v_{ij} = p_{ij} - \sum_k \sum_l t_{ik} p_{kl} t_{jl}. \tag{10}$$

Thus each element of V is a linear combination of the elements of P_0. We may therefore write

$$\text{vec}(V) = S \text{ vec}(P_0) \tag{11}$$

where S is an appropriate square matrix, whose form depends on the definition of $\text{vec}(\cdot)$. Expression (11) is a set of linear equations, from which we may obtain P_0.

We consider three definitions of $\text{vec}(A)$, where A is a given symmetric square matrix:

(1) $\text{vec}(A)$ is obtained by stacking the columns of A. In this case

$$S = I - T \otimes T; \tag{12}$$

see Harvey and Phillips (1976).

(2) $\text{vec}(A)$ is obtained by stacking the columns of the lower triangular part of A. This makes use of the symmetry of V and P_0, and reduces the problem to one of solving $r(r + 1)/2$ linear equations.

When we take into account the form of T, we see that the matrix S contains many zeros. We may therefore solve (10) by means of a series of Givens transformations of S, thus obtaining the QR decomposition of S. The matrix S is processed row by row, and the method takes into account leading zeros in the rows of S to reduce computing time. Solving the equations this way on an ICL 4130 computer for a $MA(4)$ process was faster than the standard NAG routine by a factor of about three.

A further saving in time may be made for pure moving average processes when the equation (10) forms a triangular system, and P may be found by backsubstitution.

(3) $\text{vec}(A)$ is obtained by stacking the columns of the lower triangular part of A, beginning at column 2, with the first column attached at the end. This formulation attempts to bring more leading zeros in the rows of S, leading to a reduction in the time taken to evaluate P_0. This formulation is used in subroutine *STARMA* for processes with autoregressive components. For pure moving average processes, the method described in the preceding paragraph is used. [The referee has suggested an alternative approach in which P_0 is derived from the dispersion matrix of $w_0, \ldots, w_{1-r}, \epsilon_0, \ldots, \epsilon_{1-r}$, and computed using the algorithm of McLeod (1975).]

The recursions in subroutine *KARMA* are programmed efficiently by taking account of the zero values which arise in predetermined positions in the various matrices. An important saving is effected in computing TP_tT' in the prediction recursion because of the special nature of P_t; see Harvey and Phillips (1976).

STRUCTURE

SUBROUTINE STARMA(IP, IQ, IR, NP, PHI, THETA, A, P, V, THETAB, XNEXT, XROW, RBAR, NRBAR, IFAULT)

Formal parameters

IP	Integer	input:	the value of p.
IQ	Integer	input:	the value of q.
IR	Integer	input.	the value of $r = \max$ $(p, q + 1)$.
NP	Integer	input:	the value of $r(r + 1)/2$.
PHI	Real array (IR)	input:	the value of ϕ in the first p locations.
		output:	contains the first column of T.
THETA	Real array (IR)	input:	the value of θ in the first q locations.
A	Real array (IR)	output:	on exit contains a_0.
P	Real array (NP)	output:	on exit contains P_0, stored as a lower triangular matrix, column by column.
V	Real array (NP)	output:	on exit contains RR', stored as a lower triangular matrix, column by column.

THETAB	Real array (*NP*)	workspace:	used to calculate P.
XNEXT	Real array (*NP*)	workspace:	used to calculate P.
XROW	Real array (*NP*)	workspace:	used to calculate P.
RBAR	Real array (*NRBAR*)	workspace:	used to calculate P.
NRBAR	Integer	input:	the value of $NP*(NP-1)/2$.
IFAULT	Integer	output:	a fault indicator, equal to

    1 if $IP < 0$;

    2 if $IQ < 0$;

    3 if $IP < 0$ and $IQ < 0$;

    4 if $IP = IQ = 0$;

    5 if $IR \neq MAX(IP, IQ + 1)$;

    6 if $NP \neq IR*(IR + 1)/2$;

    7 if $NRBAR \neq NP*$
$(NP - 1)/2$;

    8 if $IP = 1$ and $IQ = 0$
(Subroutine *STARMA* is
not appropriate for an
$AR(1)$ process.);

    0 otherwise.

SUBROUTINE KARMA(IP, IQ, IR, NP, PHI, THETA, A, P, V, N, W, RESID, SUMLOG, SSQ, IUPD, DELTA, E, NIT)

Formal parameters

IP	Integer	input:	the value of p.
IQ	Integer	input:	the value of q.
IR	Integer	input:	the value of $r = \max(p, q + 1)$.
NP	Integer	input:	the value of $r(r + 1)/2$.
PHI	Real array (*IR*)	input:	the first column of T.
THETA	Real array (*IR*)	input:	the value of θ in the first q locations.
A	Real array (*IR*)	input:	contains a_0.
		output:	contains a_t, where $t = t^*$.
P	Real array (*NP*)	input:	contains P_0.
		output:	contains P_t, where $t = t^*$.
V	Real array (*NP*)	input:	contains RR'.
N	Integer	input:	n, the number of observations.
W	Real array (*N*)	input:	the observations.
RESID	Real array (*N*)	output:	the corresponding standardized prediction errors.
SUMLOG	Real	input:	initial value of $\Sigma \log f_t$ (zero if no previous observations).
		output:	final value of $\Sigma \log f_t$.

SSQ	Real	input:	initial value of $\Sigma \, \bar{\nu}_t^2$ (zero if no previous observations).
		output:	final value of $\Sigma \, \bar{\nu}_t^2$.
IUPD	Integer	input:	if $IUPD = 1$ the prediction equations are by-passed for the first observation. This is necessary when the value of P_0 has been obtained from *STARMA*. In this case, $P_{1|0} = P_0$ and $a_{1|0} = a_0$ and using the prediction equations as coded in *KARMA* would lead to erroneous results. For values other than 1, the prediction equations are not by-passed.
DELTA	Real	input:	when $NIT = 0$ this parameter determines the level of approximation. Negative *DELTA* ensures that the Kalman filter is used for all observations. Otherwise the filter is performed while $f_t \geqslant 1 + \delta$, 'quick recursions' being used thereafter.
E	Real array (IR)	workspace:	used to store the last q standardized prediction errors.
NIT	Integer	input:	when set to zero see description of *DELTA* for the effect of *NIT*; for non-zero values, the 'quick recursions' are performed throughout, so that a conditional likelihood is obtained.
		output:	number of observations dealt with by the Kalman filter, i.e. t^*.

SUBROUTINE KALFOR(M, IP, IR, NP, PHI, A, P, V, WORK)

Formal parameters

M	Integer	input:	the value of m, the number of steps ahead for which predictor is required.
IP	Integer	input:	the value of p.
IR	Integer	input:	the value of r.
NP	Integer	input:	$r(r + 1)/2$.
PHI	Real array (IR)	input:	contains the first column of T, the transition matrix.
A	Real array (IR)	input:	current value of a_t.
		output:	predicted value of a_{t+m}.
P	Real array (NP)	input:	current value of P_t, stored in lower triangular form, column by column.
		output:	predicted value of P_{t+m}.
V	Real array (NP)	input:	contains RR' stored in lower triangular form, column by column.
WORK	Real array (IR)	workspace:	

Auxiliary algorithms

The subroutine *STARMA* calls the auxiliary algorithms *INCLU2* (Farebrother, 1976) and *REGRES* (Gentleman, 1974). These algorithms were originally presented as Algol 60 procedures. The following modified Fortran 66 versions of these procedures are listed after subroutine *KALFOR*:

SUBROUTINE INCLU2(NP, NRBAR, WEIGHT, XNEXT, XROW, YNEXT, D,
RBAR, THETAB, SSQERR, RECRES, IRANK, IFAULT)
SUBROUTINE REGRES(NP, NRBAR, RBAR, THETAB, BETA)

The formal parameters of these subroutines correspond to those in the original Algol procedures except that NRBAR contains the value $p(p - 1)/2$, XNEXT contains the independent variables for the current observation, and XROW is workspace. Both XNEXT and XROW are real arrays of length NP, and the values in XNEXT are unchanged by a call of INCLU2.

PRECISION

To convert the routines to double precision the following changes should be made:

(1) Change *REAL* to *DOUBLE PRECISION* in each routine;
(2) Change the real constants in the *DATA* statements to double precision versions;
(3) Change *ALOG* and *SQRT* to *DLOG* and *DSQRT* in the statement functions.

TIME

The figures in Table 2 show the number of seconds taken to compute the likelihood function of various $MA(q)$ models for set values of the MA parameters. The computations were carried out on an ICL 4130 machine at the University of Kent. The classical method of evaluating the exact likelihood function involves estimating the pre-sample residuals; see Box and Jenkins (1970, Chapter 7). Our algorithm was written to be as efficient as possible, employing what is essentially a specialization of the method described by Osborn (1977) for the MA multivariate case; see Harvey and Phillips (1976). The 'quick recursions' were *not* used in the Kalman filter algorithm. However, the Kalman filter algorithm appears to be marginally faster than the classical method for small sample sizes. Both methods of computing the exact likelihood function take significantly longer than the conditional sum of squares approximation, although the results in Table 1 suggest a considerable saving of time in the Kalman filter method when the quick recursions are used.

Table 2

Comparison of times[†] required to evaluate the likelihood for $MA(q)$ models for different sample sizes

q	n	Conditional sum of squares	Classical	Kalman filter
1	20	0·23	0·39	0·37
	40	0·39	0·61	0·65
	60	0·56	0·87	0·93
	80	0·73	1·12	1·19
	100	0·89	1·38	1·46
2	20	0·23	0·46	0·41
	60	0·58	1·14	1·16
	100	0·92	1·81	1·77
3	20	0·23	0·59	0·50
4	20	0·24	0·75	0·60

[†]Number of seconds required to execute each of the three algorithms on an ICL 4130 computer.

Although these computational comparisons are restricted to MA models, we believe that for general $ARMA$ models the conclusions concerning the performance of our Kalman filter algorithm, *vis-à-vis* algorithms based on other methods are likely to be similar. Indeed a yardstick for gauging the relative efficiency of the Kalman filter method in this context is provided by noting that the time taken to evaluate the likelihood of an $ARMA(5, 4)$ model was $1\cdot00$ seconds for $n = 20$. This figure may be compared with the corresponding figure for the $MA(4)$ process presented in Table 2. The dimensions of the matrices in the filter are exactly the same for these two processes, but the likelihood for the pure MA model is computed more rapidly because of the special features exploited in evaluating P_0.

ACKNOWLEDGEMENTS

We are grateful to the SSRC for financial support in connection with our project 'Testing for Specification Error in Econometric Models' which was carried out at the University of Kent in 1976–77. We would also like to thank the editor and referee for their comments and suggestions.

EDITORS' REMARKS

No changes to the basic structure of the algorithm have been made, just minor changes principally those associated with facilitating the change to double precision outlined above.

No Remarks on this algorithm have been published in *Applied Statistics*, but readers' attention is drawn to AS 182 (Harvey and McKenzie, 1982), AS 191 (McLeod and Holanda Sales, 1983), and AS 197 (Mélard, 1984) which offer alternative facilities which may be more appropriate for certain models.

REFERENCES

Box, G. E. P. and Jenkins, G. M. (1970) *Time Series Analysis: Forecasting and Control.* San Francisco: Holden-Day.

Farebrother, R. W. (1976) Remark AS R17. Recursive residuals – a remark on Algorithm AS 75. Basic procedures for large, sparse or weighted linear least squares problems. *Appl. Statist.*, **25**, 323–324.

Gentleman, W. M. (1974) Algorithm AS 75. Basic procedures for large, sparse or weighted linear least squares problems. *Appl. Statist.*, **23**, 448–454. (See also this book, page 130)

Harvey, A. C. and McKenzie, C. R. (1982) Algorithm AS 182. Finite sample prediction from ARIMA processes. *Appl. Statist.*, **31**, 180–187.

Harvey, A. C. and Phillips, G. D. A. (1976) *Maximum likelihood estimation of autoregressive – moving average models by Kalman filtering.* University of Kent: QSS Discussion Paper No. 38.

Harvey, A. C. and Phillips, G. D. A. (1979) Maximum likelihood estimation of regression models with autoregressive-moving average disturbances. *Biometrika*, **66**, 49–58.

McLeod, I. (1975) The derivation of the theoretical autocovariance function of autoregressive-moving average time series. *Appl. Statist.*, **24**, 255–256.

McLeod, A. I. and Holanda Sales, P. R. (1983) Algorithm AS 191. An algorithm for approximate likelihood calculation of ARMA and seasonal ARMA models. *Appl. Statist.*, **32**, 211–223.

Mélard, G. (1984) Algorithm AS 197. A fast algorithm for the exact likelihood of autoregressive − moving average models. *Appl. Statist.*, **33**, 104−114.

Osborn, D. R. (1977) Exact and approximate maximum likelihood estimators for vector moving average processes. *J. R. Statist. Soc. B*, **39**, 114−118.

```
      SUBROUTINE STARMA(IP, IQ, IR, NP, PHI, THETA, A, P, V, THETAB,
     $   XNEXT, XROW, RBAR, NRBAR, IFAULT)
C
C        ALGORITHM AS 154   APPL. STATIST. (1980) VOL.29, P.311
C
C        INVOKING THIS SUBROUTINE SETS THE VALUES OF V AND PHI, AND
C        OBTAINS THE INITIAL VALUES OF A AND P.
C        THIS ROUTINE IS NOT SUITABLE FOR USE WITH AN AR(1) PROCESS.
C        IN THIS CASE THE FOLLOWING INSTRUCTIONS SHOULD BE USED FOR
C        INITIALISATION.
C        V(1) = 1.0
C        A(1) = 0.0
C        P(1) = 1.0 / (1.0 - PHI(1) * PHI(1))
C
      REAL PHI(IR), THETA(IR), A(IR), P(NP), V(NP), THETAB(NP),
     $   XNEXT(NP), XROW(NP), RBAR(NRBAR), VJ, PHII, PHIJ, SSQERR,
     $   RECRES, YNEXT, ZERO, ONE
C
      DATA ZERO, ONE /0.0, 1.0/
C
C        CHECK FOR FAILURE INDICATION.
C
      IFAULT = 0
      IF (IP .LT. 0) IFAULT = 1
      IF (IQ .LT. 0) IFAULT = IFAULT + 2
      IF (IP .EQ. 0 .AND. IQ .EQ. 0) IFAULT = 4
      K = IQ + 1
      IF (K .LT. IP) K = IP
      IF (IR .NE. K) IFAULT = 5
      IF (NP .NE. IR * (IR + 1) / 2) IFAULT = 6
      IF (NRBAR .NE. NP * (NP - 1) / 2) IFAULT = 7
      IF (IR .EQ. 1) IFAULT = 8
      IF (IFAULT .NE. 0) RETURN
C
C        NOW SET A(0), V AND PHI.
C
      DO 10 I = 2, IR
      A(I) = ZERO
      IF (I .GT. IP) PHI(I) = ZERO
      V(I) = ZERO
      IF (I .LE. IQ + 1) V(I) = THETA(I - 1)
   10 CONTINUE
      A(1) = ZERO
      IF (IP .EQ. 0) PHI(1) = ZERO
      V(1) = ONE
      IND = IR
      DO 20 J = 2, IR
      VJ = V(J)
      DO 20 I = J, IR
      IND = IND + 1
      V(IND) = V(I) * VJ
   20 CONTINUE
C
C        NOW FIND P(0).
C
      IF (IP .EQ. 0) GOTO 300
```

```
C
C            THE SET OF EQUATIONS S * VEC(P(0)) = VEC(V)
C            IS SOLVED FOR VEC(P(0)).
C            S IS GENERATED ROW BY ROW IN THE ARRAY XNEXT.
C            THE ORDER OF ELEMENTS IN P IS CHANGED, SO AS TO
C            BRING MORE LEADING ZEROS INTO THE ROWS OF S,
C            HENCE ACHIEVING A REDUCTION OF COMPUTING TIME.
C
         IRANK = 0
         SSQERR = ZERO
         DO 40 I = 1, NRBAR
   40 RBAR(I) = ZERO
         DO 50 I = 1, NP
         P(I) = ZERO
         THETAB(I) = ZERO
         XNEXT(I) = ZERO
   50 CONTINUE
         IND = 0
         IND1 = 0
         NPR = NP - IR
         NPR1 = NPR + 1
         INDJ = NPR1
         IND2 = NPR
         DO 110 J = 1, IR
         PHIJ = PHI(J)
         XNEXT(INDJ) = ZERO
         INDJ = INDJ + 1
         INDI = NPR1 + J
         DO 110 I = J, IR
         IND = IND + 1
         YNEXT = V(IND)
         PHII = PHI(I)
         IF (J .EQ. IR) GOTO 100
         XNEXT(INDJ) = -PHII
         IF (I .EQ. IR) GOTO 100
         XNEXT(INDI) = XNEXT(INDI) - PHIJ
         IND1 = IND1 + 1
         XNEXT(IND1) = -ONE
  100 XNEXT(NPR1) = -PHII * PHIJ
         IND2 = IND2 + 1
         IF (IND2 .GT. NP) IND2 = 1
         XNEXT(IND2) = XNEXT(IND2) + ONE
         CALL INCLU2(NP, NRBAR, ONE, XNEXT, XROW, YNEXT,
     $     P, RBAR, THETAB, SSQERR, RECRES, IRANK, IFAIL)
C
C            NO NEED TO CHECK IFAIL AS WEIGHT = 1.0
C
         XNEXT(IND2) = ZERO
         IF (I .EQ. IR) GOTO 110
         XNEXT(INDI) = ZERO
         INDI = INDI + 1
         XNEXT(IND1) = ZERO
  110 CONTINUE
         CALL REGRES(NP, NRBAR, RBAR, THETAB, P)
C
C            NOW RE-ORDER P.
C
         IND = NPR
         DO 200 I = 1, IR
         IND = IND + 1
         XNEXT(I) = P(IND)
  200 CONTINUE
         IND = NP
         IND1 = NPR
         DO 210 I = 1, NPR
         P(IND) = P(IND1)
         IND = IND - 1
         IND1 = IND1 - 1
```

```
  210 CONTINUE
      DO 220 I = 1, IR
  220 P(I) = XNEXT(I)
      RETURN
C
C         P(0) IS OBTAINED BY BACKSUBSTITUTION FOR
C         A MOVING AVERAGE PROCESS.
C
  300 INDN = NP + 1
      IND = NP + 1
      DO 310 I = 1, IR
      DO 310 J = 1, I
      IND = IND - 1
      P(IND) = V(IND)
      IF (J .EQ. 1) GOTO 310
      INDN = INDN - 1
      P(IND) = P(IND) + P(INDN)
  310 CONTINUE
      RETURN
      END
C
      SUBROUTINE KARMA(IP, IQ, IR, NP, PHI, THETA, A, P,
     $ V, N, W, RESID, SUMLOG, SSQ, IUPD, DELTA, E, NIT)
C
C         ALGORITHM AS 154.1  APPL. STATIST. (1980) VOL.29, P.311
C
C         INVOKING THIS SUBROUTINE UPDATES A, P, SUMLOG AND SSQ BY
C         INCLUSION OF DATA VALUES W(1) TO W(N). THE CORRESPONDING
C         VALUES OF RESID ARE ALSO OBTAINED.
C         WHEN FT IS LESS THAN (1 + DELTA), QUICK RECURSIONS ARE USED.
C
      REAL PHI(IR), THETA(IR), A(IR), P(NP), V(NP), W(N), RESID(N),
     $ E(IR), SUMLOG, SSQ, DELTA, WNEXT, A1, DT, ET, FT, UT, G,
     $ ZERO, ZLOG, ZSQRT
C
      DATA ZERO /0.0/
C
      ZLOG(G) = ALOG(G)
      ZSQRT(G) = SQRT(G)
C
      IR1 = IR - 1
      DO 10 I = 1, IR
   10 E(I) = ZERO
      INDE = 1
C
C         FOR NON-ZERO VALUES OF NIT, PERFORM QUICK RECURSIONS.
C
      IF (NIT .NE. 0) GOTO 600
      DO 500 I = 1, N
      WNEXT = W(I)
C
C         PREDICTION.
C
      IF (IUPD .EQ. 1 .AND. I .EQ. 1) GOTO 300
C
C         HERE DT = FT - 1.0
C
      DT = ZERO
      IF (IR .NE. 1) DT = P(IR + 1)
      IF (DT .LT. DELTA) GOTO 610
      A1 = A(1)
      IF (IR .EQ. 1) GOTO 110
      DO 100 J = 1, IR1
  100 A(J) = A(J + 1)
  110 A(IR) = ZERO
      IF (IP .EQ. 0) GOTO 200
      DO 120 J = 1, IP
  120 A(J) = A(J) + PHI(J) * A1
```

```
  200 IND = 0
      INDN = IR
      DO 210 L = 1, IR
      DO 210 J = L, IR
      IND = IND + 1
      P(IND) = V(IND)
      IF (J .EQ. IR) GOTO 210
      INDN = INDN + 1
      P(IND) = P(IND) + P(INDN)
  210 CONTINUE
C
C         UPDATING.
C
  300 FT = P(1)
      UT = WNEXT - A(1)
      IF (IR .EQ. 1) GOTO 410
      IND = IR
      DO 400 J = 2, IR
      G = P(J) / FT
      A(J) = A(J) + G * UT
      DO 400 L = J, IR
      IND = IND + 1
      P(IND) = P(IND) - G * P(L)
  400 CONTINUE
  410 A(1) = WNEXT
      DO 420 L = 1, IR
  420 P(L) = ZERO
      RESID(I) = UT / ZSQRT(FT)
      E(INDE) = RESID(I)
      INDE = INDE + 1
      IF (INDE .GT. IQ) INDE = 1
      SSQ = SSQ + UT * UT / FT
      SUMLOG = SUMLOG + ZLOG(FT)
  500 CONTINUE
      NIT = N
      RETURN
C
C         QUICK RECURSIONS
C
  600 I = 1
  610 NIT = I - 1
      DO 650 II = I, N
      ET = W(II)
      INDW = II
      IF (IP .EQ. 0) GOTO 630
      DO 620 J = 1, IP
      INDW = INDW - 1
      IF (INDW .LT. 1) GOTO 630
      ET = ET - PHI(J) * W(INDW)
  620 CONTINUE
  630 IF (IQ .EQ. 0) GOTO 645
      DO 640 J = 1, IQ
      INDE = INDE - 1
      IF (INDE .EQ. 0) INDE = IQ
      ET = ET - THETA(J) * E(INDE)
  640 CONTINUE
  645 E(INDE) = ET
      RESID(II) = ET
      SSQ = SSQ + ET * ET
      INDE = INDE + 1
      IF (INDE .GT. IQ) INDE = 1
  650 CONTINUE
      RETURN
      END
C
      SUBROUTINE KALFOR(M, IP, IR, NP, PHI, A, P, V, WORK)
C
C         ALGORITHM AS 154.2  APPL. STATIST. (1980) VOL.29, P.311
C
```

```
C          INVOKING THIS SUBROUTINE OBTAINS PREDICTIONS
C          OF A AND P, M STEPS AHEAD.
C
      REAL PHI(IR), A(IR), P(NP), V(NP), WORK(IR), DT,
     $ A1, PHII, PHIJ, PHIJDT, ZERO
C
      DATA ZERO /0.0/
C
      IR1 = IR - 1
      DO 300 L = 1, M
C
C          PREDICT A.
C
      A1 = A(1)
      IF (IR .EQ. 1) GOTO 110
      DO 100 I = 1, IR1
  100 A(I) = A(I + 1)
  110 A(IR) = ZERO
      IF (IP .EQ. 0) GOTO 200
      DO 120 J = 1, IP
  120 A(J) = A(J) + PHI(J) * A1
C
C          PREDICT P.
C
  200 DO 210 I = 1, IR
  210 WORK(I) = P(I)
      IND = 0
      IND1 = IR
      DT = P(1)
      DO 220 J = 1, IR
      PHIJ = PHI(J)
      PHIJDT = PHIJ * DT
      DO 220 I = J, IR
      IND = IND + 1
      PHII = PHI(I)
      P(IND) = V(IND) + PHII * PHIJDT
      IF (J .LT. IR) P(IND) = P(IND) + WORK(J + 1) * PHII
      IF (I .EQ. IR) GOTO 220
      IND1 = IND1 + 1
      P(IND) = P(IND) + WORK(I + 1) * PHIJ + P(IND1)
  220 CONTINUE
  300 CONTINUE
      RETURN
      END
C
      SUBROUTINE INCLU2(NP, NRBAR, WEIGHT, XNEXT, XROW, YNEXT, D, RBAR,
     $ THETAB, SSQERR, RECRES, IRANK, IFAULT)
C
C          ALGORITHM AS 154.3  APPL. STATIST. (1980) VOL.29, P.311
C
C          FORTRAN VERSION OF REVISED VERSION OF ALGORITHM AS 75.1
C          APPL. STATIST. (1974) VOL.23, P.448
C          SEE REMARK AS R17 APPL. STATIST. (1976) VOL.25, P.323
C
      REAL XNEXT(NP), XROW(NP), D(NP), RBAR(NRBAR), THETAB(NP),
     $ WEIGHT, YNEXT, SSQERR, RECRES, WT, Y, DI, DPI, XI, XK,
     $ CBAR, SBAR, RBTHIS, ZERO, ZSQRT
C
      DATA ZERO /0.0/
C
      ZSQRT(Y) = SQRT(Y)
C
C          INVOKING THIS SUBROUTINE UPDATES D, RBAR, THETAB, SSQERR
C          AND IRANK BY THE INCLUSION OF XNEXT AND YNEXT WITH A
C          SPECIFIED WEIGHT. THE VALUES OF XNEXT, YNEXT AND WEIGHT WILL
C          BE CONSERVED. THE CORRESPONDING VALUE OF RECRES IS CALCULATED.
C
      Y = YNEXT
      WT = WEIGHT
```

```
      DO 10 I = 1, NP
   10 XROW(I) = XNEXT(I)
      RECRES = ZERO
      IFAULT = 1
      IF (WT .LE. ZERO) RETURN
      IFAULT = 0
C
      ITHISR = 0
      DO 50 I = 1, NP
      IF (XROW(I) .NE. ZERO) GOTO 20
      ITHISR = ITHISR + NP - I
      GOTO 50
   20 XI = XROW(I)
      DI = D(I)
      DPI = DI + WT * XI * XI
      D(I) = DPI
      CBAR = DI / DPI
      SBAR = WT * XI / DPI
      WT = CBAR * WT
      IF (I .EQ. NP) GOTO 40
      I1 = I + 1
      DO 30 K = I1, NP
      ITHISR = ITHISR + 1
      XK = XROW(K)
      RBTHIS = RBAR(ITHISR)
      XROW(K) = XK - XI * RBTHIS
      RBAR(ITHISR) = CBAR * RBTHIS + SBAR * XK
   30 CONTINUE
   40 XK = Y
      Y = XK - XI * THETAB(I)
      THETAB(I) = CBAR * THETAB(I) + SBAR * XK
      IF (DI .EQ. ZERO) GOTO 100
   50 CONTINUE
      SSQERR = SSQERR + WT * Y * Y
      RECRES = Y * ZSQRT(WT)
      RETURN
  100 IRANK = IRANK + 1
      RETURN
      END
C
      SUBROUTINE REGRES(NP, NRBAR, RBAR, THETAB, BETA)
C
C         ALGORITHM AS 154.4   APPL. STATIST. (1980) VOL.29, P.311
C
C         REVISED VERSION OF ALGORITHM AS 75.4
C         APPL. STATIST. (1974) VOL.23, P.448
C         INVOKING THIS SUBROUTINE OBTAINS BETA BY BACKSUBSTITUTION
C         IN THE TRIANGULAR SYSTEM RBAR AND THETAB.
C
      REAL RBAR(NRBAR), THETAB(NP), BETA(NP), BI
      ITHISR = NRBAR
      IM = NP
      DO 50 I = 1, NP
      BI = THETAB(IM)
      IF (IM .EQ. NP) GOTO 30
      I1 = I - 1
      JM = NP
      DO 10 J = 1, I1
      BI = BI - RBAR(ITHISR) * BETA(JM)
      ITHISR = ITHISR - 1
      JM = JM - 1
   10 CONTINUE
   30 BETA(IM) = BI
      IM = IM - 1
   50 CONTINUE
      RETURN
      END
```

Algorithm AS 168

SCALE SELECTION AND FORMATTING
By W. Douglas Stirling
Queen Mary College, Mile End Road, London E1 4NS, UK

Present address: Department of Maths and Statistics, Massey University, Palmerston North, New Zealand.

Keywords: Scale; Plot; Formatting.

LANGUAGE
Fortran 66 and 77.

DESCRIPTION AND PURPOSE

The algorithm is in two separate sections. Subroutine *SCALE* chooses a scale to cover given minimum and maximum values such that the values to be printed at specified regular intervals on the scale are as 'neat' as possible. Values are chosen with less significant figures than previous algorithms, especially when the values are printed at intervals other than every 5th or 10th plotting position on the scale. The algorithm can therefore be successfully used with special-purpose graphical devices (with the order of 1000 plotting positions) as well as character-output devices.

If the values labelling a scale (such as those chosen by *SCALE*) are to be printed as neatly as possible in Fortran, variable format output must be used. Subroutine *AXIS* finds a suitable format for printing in these circumstances, when told the size of field available for each value.

Choosing neat values
The present subroutine *SCALE* is based on a previous scaling algorithm AS 96 (Nelder, 1976), but differs from it in several respects.

AS 96 concentrates on finding a neat step between adjacent plotting positions and makes the first plotting position a multiple of these steps from the origin. (A subsequent shift may be performed if the first or last plotting position can be set to zero). If values are printed other than every 1, 10 or 100 plotting positions, the steps between printed values can have three or more significant figures. Another characteristic of the approach is that even with 10 plotting positions per value, that printed at the lowest plotting position is often not a multiple of 10 steps from the origin, resulting in printed values of, for example, $14\cdot4, 26\cdot4, 38\cdot4, \ldots$.

The algorithm being described here makes use of the number of plotting positions per printed value, *MPV*. It ensures that the step between *printed* values is neat (no more than two significant digits) and that the printed values themselves have no digits less significant than these.

SCALE first finds the smallest neat step between printed values, *STEP* (with significant digits 10, 12, 15, 20, 25, 30, 40, 50, 60 or 80), such that the scale's range will be larger than the range of the values to be covered by the scale. If this scale cannot be shifted to make the lowest printed value a multiple of *STEP*, while still including the minimum and maximum values to be covered, *FMN* and *FMX*, a choice is made between increasing *STEP* or making the lowest printed value a multiple of *STEP/K* where *K* is an integer and *STEP/K* has no less significant digits than *STEP*'s two. The latter strategy is chosen with increasing *K* until failure with some *K* could only have occurred with (*FMX−FMN*) big enough to cover at least 70 per cent of a scale using the next larger *STEP*.

Although neat values are always chosen for printing, this approach also has two disadvantages. Firstly with some values of *MPV* (e.g. 7) the step between plotting positions may not be neat, making interpolation difficult. *MPV* can, however, be set to 5 or 10 if this is thought to be important, though in practice accurate estimates of intermediate points are rarely required. Also the data can only be guaranteed to cover 70 per cent of the scale, whereas AS 96 guarantees covering 75 per cent. Both algorithms, however, normally perform much better than this and the cases where this algorithm is worst are generally those for which AS 96 produces unsatisfactory scales.

Three other more minor improvements over AS 96 are given below:

(1) The test for 'effective equality' of *FMN* and *FMX* is not symmetric in AS 96 (e.g. ranges 0 to 10^{-6} and -10^{-6} to 0 are treated differently). It is defined here when

$$(FMX - FMN) \leqslant TOL * \max(|FMX|, |FMN|).$$

The scales chosen then cover *FMN* and *FMX* but do not separate them.

(2) Integer overflow can occur with AS 96 on computers with 16-bit integers if *FMN* is a large number of 'neat steps' from the origin (e.g. range 1000 to 1001 with 100 plotting positions). Real arithmetic is used at the corresponding point here.

(3) The value *FMX* itself may not be mapped onto the scale in AS 96 (e.g. to cover the range 0·5 to 10·5 with 11 plotting positions, values 0, 1, 2,..., 10 are chosen by AS 96, whereas 10·5 may be rounded to 11). Values up to *BIAS* * (*FMX−FMN*) larger than *FMX* and smaller than *FMN* are also mapped onto this algorithm's scales.

Finding a good printing format for the scale
After the minimum value and step size have been chosen for a scale (e.g. by *SCALE*), the scale values usually have to be printed. Use of fixed *G* format (the most flexible in standard Fortran) results in some numbers unnecessarily printed with an exponent (e.g. values less than 0·1) and a possible loss of varying digits by truncation when the field size is restricted. Subroutine *AXIS* determines how many digits to print after the decimal point, *IRPRIN*, when a variable *F* format allowing *MAXPR* printed digits is used.

When all significant digits cannot be printed in this way, *AXIS* multiplies the scale values by 10^{-IFACT} and returns *IFACT* as a separate scaling factor for the whole scale. When even this means that varying digits would be lost on printing, the most significant non-varying ones are removed as an offset, *OFFSET*, for the whole scale.

In all cases the values to be printed are returned in array *VALS*, and should be printed with variable $F(MAXPR + 2)$. $(IRPRIN)$ format. The values represented are

$$(VALS(I) + OFFSET) * 10^{IFACT}.$$

STRUCTURE

SUBROUTINE SCALE (FMN, FMX, N, MPV, VALMIN, STEP, NVALS, IR, IFAULT)

Formal parameters

FMN	Real	input:	minimum value to be included in scale.
FMX	Real	input:	maximum value to be included in scale.
N	Integer	input:	number of distinct plotting positions.
MPV	Integer	input:	number of plotting positions per printed value.
VALMIN	Real	output:	neat value for lowest plotting position.
STEP	Real	output:	neat step between printed values.
NVALS	Integer	output:	number of values to be printed.
IR	Integer	output:	number of left-shifts required to make all printed values integers.
IFAULT	Integer	output:	fault indicator.

Fault indicator

$IFAULT = -1$ *FMN* and *FMX* effectively equal (non-fatal);

   0 no fault;

   1 illegal range $(FMN > FMX)$;

   2 illegal number of plotting positions $(N < 2$ or $N > 10,000)$;

   4 illegal $MPV (MPV \leqslant 0$ or $MPV \geqslant N)$;

   otherwise the sum of the values above for those fatal faults detected.

Constants

Real value *TOL* defines effective equality between *FMN* and *FMX*. Real value

BIAS defines the points outside *FMN* to *FMX* that must also be included in the scale. Both are fully described earlier.

SUBROUTINE AXIS (VALMIN, STEP, NVALS, MAXPR, IR, IRPRIN, OFFSET, IFACT, VALS, IV, IFAULT)

Formal parameters

VALMIN	Real	input:	lowest value on scale.
STEP	Real	input:	step between values on scale.
NVALS	Integer	input:	number of values on scale.
MAXPR	Integer	input:	maximum number of digits that can be printed in each value (requiring *MAXPR* + 2 print positions to allow for the sign and decimal point).
IR	Integer	input:	number of left shifts required to make all printed values integers.
IRPRIN	Integer	output:	number of digits required after the decimal point on printing.
OFFSET	Real	output:	offset for the scale (see Description and Purpose).
IFACT	Integer	output:	power of ten by which the printed values should be multiplied (see Description and Purpose).
VALS	Real array (*IV*)	output:	the values to be printed occupy the first *NVALS* elements.
IV	Integer	input:	dimension of *VALS* (*IV* ⩾ *NVALS*).
IFAULT	Integer	output:	fault indicator.

Fault indicator

$IFAULT =$ 0 no fault;

1 less than 2 values to be printed;

2 step size not positive;

4 improper digits allowed for printing ($MAXPR < 2$ or $MAXPR > MPRMAX$);

8 *VALS* not big enough for values ($NVALS > IV$);

16 *IR* inconsistent with *VALMIN* and *STEP* or too large ($IR > IRMAX$);

otherwise the sum of the values above for those faults detected.

Constants

MPRMAX denotes the maximum number of digits that *AXIS* will accept as reasonable for printing. *IRMAX* is the maximum number of decimal places that are acceptable in scale values. Both constants may be increased if necessary.

PRECISION

In normal circumstances single precision should be adequate. However, it is possible to convert the routines to double precision by:

(1) Changing *REAL* to *DOUBLE PRECISION* in the declarations;
(2) Changing the real constants in the *DATA* statements to double precision versions;
(3) Changing the right-hand sides of the statement functions into suitable double precision equivalents (but note that under standard Fortran 66 there is no equivalent to *AINT* or *FLOAT,* though many compilers offer such routines; in their absence replace *AINT(S)* by *S–DMOD(S,ONE)* and *FLOAT(I)* by *DBLE(FLOAT(I))*).

EDITORS' REMARKS

The only changes made are those facilitating the change to double precision outlined above.

REFERENCE

Nelder, J. A. (1976) Algorithm AS 96. A simple algorithm for scaling graphs. *Appl. Statist.*, **25**, 94–96.


```
      SUBROUTINE SCALE(FMN, FMX, N, MPV, VALMIN, STEP, NVALS,
     $ IR, IFAULT)
C
C        ALGORITHM AS 168  APPL. STATIST. (1981) VOL.30, P.339
C
C        CALCULATES NEAT VALUES FOR LOWEST PRINT POSITION AND
C        STEP BETWEEN PRINTED VALUES ON A SCALE TO INCLUDE FMN
C        AND FMX
C
      REAL UNIT(12), FMN, FMX, VALMIN, STEP, FMAX, FMIN, FINTER, AJ,
     $ S, TEMP, TSTEP, ZABS, ZFLOAT, ZINT, ZMAX1
      REAL TOL, BIAS, COVER, ZERO, TENTH, HALF, ONE, TWO, TEN, HUND
C
      DATA NUNIT /12/
      DATA UNIT(1), UNIT(2), UNIT(3),  UNIT(4),  UNIT(5),  UNIT(6),
     $    UNIT(7), UNIT(8), UNIT(9), UNIT(10), UNIT(11), UNIT(12)
     $    /12.0,    15.0,    20.0,     25.0,     30.0,     40.0,
     $     50.0,    60.0,    80.0,    100.0,    120.0,    150.0/
      DATA TOL, BIAS /5.0E-6, 1.0E-5/
      DATA MINN, MAXN, COVER /2, 10000, 0.7/
      DATA ZERO, TENTH, HALF, ONE, TWO,  TEN,  HUND
     $    /0.0,   0.1,   0.5, 1.0, 2.0, 10.0, 100.0/
```

```
C
      ZABS(S) = ABS(S)
      ZFLOAT(I) = FLOAT(I)
      ZINT(S) = AINT(S)
      ZMAX1(AJ, S) = AMAX1(AJ, S)
C
      FMAX = FMX
      FMIN = FMN
C
C         TEST FOR VALID PARAMETERS
C
      IFAULT = 0
      IF (FMAX .LT. FMIN) IFAULT = IFAULT + 1
      IF (N .LT. MINN .OR. N .GT. MAXN) IFAULT = IFAULT + 2
      IF (MPV .LE. 0 .OR. MPV .GE. N) IFAULT = IFAULT + 4
      IF (IFAULT .NE. 0) RETURN
      NVALS = (N - 1) / MPV + 1
C
C         TEST FOR VALUES EFFECTIVELY EQUAL
C
      IF (FMAX - FMIN .GT. TOL * ZMAX1(ZABS(FMAX), ZABS(FMIN)))
     $   GOTO 20
      IFAULT = -1
      IF (FMAX) 5, 10, 15
    5 FMAX = ZERO
      GOTO 20
   10 FMAX = ONE
      GOTO 20
   15 FMIN = ZERO
C
C         FIND NEAT TRIAL STEP SIZE
C
   20 FINTER = ZFLOAT(N) / ZFLOAT(MPV)
      S = (FMAX - FMIN) * (ONE + TWO * BIAS) / FINTER
      IR = 0
   25 IF (S .GT. TEN) GOTO 30
      S = S * TEN
      IR = IR + 1
      GOTO 25
   30 IF (S .LE. HUND) GOTO 35
      S = S / TEN
      IR = IR - 1
      GOTO 30
   35 DO 40 I = 1, NUNIT
      IF (S .LE. UNIT(I)) GOTO 45
   40 CONTINUE
   45 STEP = TEN ** (-IR) * UNIT(I)
C
C         FIND NEAT TRIAL START VALUE
C
      AJ = ZERO
   50 AJ = AJ + ONE
      IF (UNIT(I) - TENTH .GT. ZINT((UNIT(I) + TENTH) / AJ) * AJ)
     $   GOTO 50
      TSTEP = STEP / AJ
      TEMP = FMIN / TSTEP + AJ * (HALF / ZFLOAT(MPV) - FINTER * BIAS)
      VALMIN = ZINT(TEMP) * TSTEP
      IF (TEMP .LT. ZERO .AND. TEMP .NE. ZINT(TEMP))
     $   VALMIN = VALMIN - TSTEP
C
C         TEST WHETHER FMAX IS IN SCALE
C
      IF (FMAX .LT. VALMIN + STEP *
     $   (FINTER * (ONE - BIAS) - HALF / ZFLOAT(MPV))) GOTO 55
C
C         TRY NEW STEP OR START VALUE
C
      IF (UNIT(I) / UNIT(I + 1) * (ONE - ONE / (AJ * FINTER)) .LT.
     $   COVER) GOTO 50
```

```
        I = I + 1
        GOTO 45
C
C        GET POSITION OF LEAST SIGNIFICANT FIGURE ON SCALE
C
   55 DO 60 J = 1, 2
        AJ = AJ * TEN
        IF (UNIT(I) - TENTH .LT. ZINT((UNIT(I) + TENTH) / AJ) * AJ)
     $    IR = IR - 1
   60 CONTINUE
        RETURN
        END
C
        SUBROUTINE AXIS(VALMIN, STEP, NVALS, MAXPR, IR, IRPRIN, OFFSET,
     $    IFACT, VALS, IV, IFAULT)
C
C        ALGORITHM AS 168.1  APPL. STATIST. (1981) VOL.30, P.339
C
C        SETS UP VALUES AND FORMATS FOR PRINTING ON AN AXIS
C
        REAL VALS(IV), FMAX, VALMIN, STEP, OFFSET, FL, FS, TMAX, VMIN,
     $    VSTEP, ZERO, TENTH, HALF, ONE, TEN, ZABS, ZFLOAT, ZINT, ZMOD
C
        DATA ZERO, TENTH, HALF, ONE,   TEN
     $      /0.0,   0.1,  0.5, 1.0, 10.0/
        DATA IRMAX, MPRMAX /20, 20/
C
        ZABS(FL) = ABS(FL)
        ZFLOAT(I) = FLOAT(I)
        ZINT(FL) = AINT(FL)
        ZMOD(FL, FS) = AMOD(FL, FS)
C
C        CHECK FOR VALID PARAMETERS
C
        IFAULT = 0
        IF (NVALS .LT. 2) IFAULT = IFAULT + 1
        FMAX = VALMIN + STEP * ZFLOAT(NVALS - 1)
        IF (NVALS .GE. 2 .AND. FMAX .LE. VALMIN) IFAULT = IFAULT + 2
        IF (MAXPR .LT. 2 .OR. MAXPR .GT. MPRMAX) IFAULT = IFAULT + 4
        IF (NVALS .GT. IV) IFAULT = IFAULT + 8
        IF (IR .GT. IRMAX) IFAULT = IFAULT + 16
        IF (IFAULT .GT. 0) RETURN
C
C        FIND POSITION OF MOST SIGNIFICANT DIGIT (IL) AND NUMBER OF
C        SIGNIFICANT DIGITS OVERALL (IS) AND VARYING (IT)
C
        TMAX = TEN ** MAXPR
        FL = ZABS(FMAX)
        FS = ZABS(VALMIN)
        IL = 0
   10 IF (FL .LT. ONE .AND. FS .LT. ONE) GOTO 20
        FL = FL / TEN
        FS = FS / TEN
        IL = IL + 1
        GOTO 10
C
   20 IF (FL .GE. TENTH .OR. FS .GE. TENTH) GOTO 30
        FL = FL * TEN
        FS = FS * TEN
        IL = IL - 1
        GOTO 20
C
   30 IS = IL + IR
        IT = IS
        IF (VALMIN .LE. ZERO .AND. FMAX .GE. ZERO) GOTO 50
   40 FL = ZMOD(FL, ONE) * TEN
        FS = ZMOD(FS, ONE) * TEN
        IF (IT .LE. 0) GOTO 1016
        IF (INT(FL) .NE. INT(FS)) GOTO 50
```

```
      IT = IT - 1
      GOTO 40
C
C         DECIDE ON PRINTING FORMAT
C
   50 IFACT = 0
      OFFSET = ZERO
      IRPRIN = MAXO(IR, 0)
      ILPRIN = MAXO(IL, 0)
C
      IF (IRPRIN + ILPRIN .LE. MAXPR) GOTO 70
      IF (IS .LE. MAXPR) GOTO 60
      IRPRIN = MAXPR - 1
      IFACT = MAXO(IT, MAXPR) - 1 - IR
      GOTO 70
   60 IFACT = IL - 1
      IRPRIN = IS - 1
   70 FS = TEN ** (-IFACT)
      VSTEP = STEP * FS
      VMIN = VALMIN * FS
      IF (IS .LE. MAXPR) GOTO 80
      OFFSET = ZINT(VMIN / TEN) * TEN
      VMIN = VMIN - OFFSET
C
C         WRITE VALUES FOR AXIS
C
   80 DO 90 I = 1, NVALS
      VALS(I) = VMIN
      VMIN = VMIN + VSTEP
   90 CONTINUE
C
C         CHECK THAT ALL VALUES CAN BE PRINTED
C
      FS = TENTH ** IRPRIN
      IF (ZABS(VALS(1)) * FS + HALF .LT. TMAX .AND.
     $ ZABS(VALS(NVALS)) * FS + HALF .LT. TMAX) RETURN
      IL = IL + 1
      IS = IS + 1
      IT = IT + 1
      GOTO 50
C
C         ERROR INDICATOR
C
 1016 IFAULT = 16
      RETURN
      END
```

Algorithm AS 169

AN IMPROVED ALGORITHM FOR SCATTER PLOTS
By W. Douglas Stirling
Queen Mary College, Mile End Road, London E1 4NS, UK

Present address: Department of Maths and Statistics, Massey University, Palmerston North, New Zealand.

Keywords: Graph; Scatter plot.

LANGUAGE

Fortran 66.

DESCRIPTION AND PURPOSE

The algorithm produces a scatter plot of one variable on a horizontal axis against several variables on a vertical axis. It owes much to the plotting algorithm AS 44 (Sparks, 1971) and is structured similarly to it, but is an improvement in several respects discussed below.

Values on the axis

Two external subroutines are used to obtain a suitable scale for each axis. SCALE must find neat values to label an axis which covers a given range of data. AXIS chooses a format for printing these values, possibly with a scaling factor (a power of ten) and an offset which are separately printed for each axis. AS 168 (Stirling, 1981) provides suitable versions of these subroutines.

Values are printed every 10th column on the horizontal axis and every 5th line on the vertical axis, whereas AS 44 puts values at every 10th position on both. Variable formats using the output of *AXIS* print only the significant digits of these numbers and avoid the use of exponents. (The fixed *G* format used in AS 44 often resorts to printing exponents unnecessarily and can cause a different format for different values on the same axis.) Many of the characters in the formatting arrays *IFORM1* and *IFORM2* could be packed more in specific machines.

User-specified scales

AS 44 uses an index *IND* to specify which scales are to be calculated from the data. The present algorithm uses the given minimum and maximum (from *SCALEX* or *SCALEY*) as fixed if minimum < maximum, otherwise they are derived from the data. Data points are omitted from the plot if they do not fit in such user-specified ranges, and the number of such points is returned as a non-fatal error code.

Whereas AS 44 does no neatening of user-specified scales, these are also neatened here.

Size of plot

Any sizes of plot can be produced, whereas AS 44 rounds sizes to multiples of 10. The maximum height and width of the 'page' used for the plot is limited to *MAXHT* and *MAXWD* (set to 50 rows and 132 columns here). These values can be easily altered with a maximum of 100 rows and 160 columns. The corresponding minimum sizes for the plot are 6 rows and 11 columns, excluding the scales, and plot sizes requested outside these limits are treated as being the appropriate maximum or minimum possible.

Marks made

The algorithm uses marks '*', '0', '+', '*X*', '=' to represent different vertical variables (this list can be added to if required provided *MAXY* is set to the number of marks available). However, whereas AS 44 uses a special character '0' to represent multiple values, '2', '3',..., '9' are used here to give the number of points as suggested by Baptista and Casagrande (unpublished) ('9' represents 9 or more).

A further modification is made if parameter *ISTAND* = 0. In this case, the different vertical variables are not distinguished. However, an attempt is made to get better resolution on the vertical scale, with ',' and '·' being used to distinguish between values with lower and higher values in each step. If 2 values fall in one step ',' is printed (even if both values are in same half of step). Three or more values are shown as '3', '4',..., '9' as before. This is particularly useful to show gradual curves and high densities of points.

It should be noted that the two characters '·' and ',' are not in the standard Fortran character set; if they are not available on a particular computer, alternative symbols should be substituted in the *DATA* statements for variables *IAPOST* and *ISEMI*. Similarly an alternative might have to be found for ':' which occurs in the *DATA* statement for *ICOLON* and in *FORMAT* labelled 1.

Time and store

The largest part of the data space used will normally be the original $N \times M$ array of data.

Similarly, the largest part of the time taken will usually be in scanning the data to produce each line of the plot. This will be $O(LN)$ if there are L lines in the plot and N points.

STRUCTURE

SUBROUTINE SCATPL(A, N, M, ICY, NCY, ICX, IWRITE, NY, NX, SCALEY, SCALEX, ISTAND, IFAULT)

Formal parameters

A	Real array (*N*, *M*)	input:	the data are in this array, each variate occupying one column.

N	Integer	input:	number of rows of A.
M	Integer	input:	number of columns of A.
ICY	Integer array (NCY)	input:	contains the numbers of the columns containing Y-variates.
NCY	Integer	input:	number of Y-variates.
ICX	Integer	input:	column number of the X-variate.
$IWRITE$	Integer	input:	the channel number for output.
NY	Integer	input:	number of lines for Y-axis.
NX	Integer	input:	number of characters for X-axis.
$SCALEY$	Real array (2)	input:	lower and upper limits for Y-axis. If lower $\geq$ upper these are calculated from data.
$SCALEX$	Real array (2)	input:	lower and upper limits for X-axis. If lower $\geq$ upper these are calculated from data.
$ISTAND$	Integer	input:	0 — modified plot with extra resolution in Y-axis; otherwise — standard plot.
$IFAULT$	Integer	output:	fault indicator.

Fault indicator

$IFAULT$ < 0 negative of number of points that could not be plotted with user-specified scales (non-fatal).

      = 0 no fault;

        1 illegal number of rows for A ($N < 1$);

        2 illegal number of columns for A ($M < 2$);

        4 illegal X variable specified ($ICX < 1$ or $ICX > M$);

        8 illegal number of Y variables ($NCY < 1$ or $NCY > 5$);

        16 illegal Y variable specified;

        32 error in $SCALE$ or $AXIS$ subroutines when used for X-axis;

        64 error in $SCALE$ or $AXIS$ subroutines when used for Y-axis;

        otherwise the sum of the values above for those fatal faults detected.

AUXILIARY ALGORITHMS

Subroutine $SCALE$ *(FMN, FMX, N, MPV, VALMIN, STEP, NVALS, IR, IFAULT)* calculates a minimum printed value, $VALMIN$, step between printed values, $STEP$, number of printed values, $NVALS$, and minimum number of left shifts to make all printed values integers, IR, for a scale with N plotting positions and values printed every MPV of these, which covers the data values FMN and FMX.

Subroutine *AXIS (VALMIN, STEP, NVALS, MAXPR, IR, IRPRIN, OFFSET, IFACT, VALS, IV, IFAULT)* suggests a printing format for a scale with *NVALS* values from *VALMIN* in steps of *STEP* where *MAXPR* significant digits can be printed, and *IR* left shifts would make all printed values integers. The values to be printed, which are returned in array *VALS*, should use format $F(MAXPR + 2) . (IRPRIN)$ and represent values $(VALS(I) + OFFSET) \times 10^{IRPRIN}$. AS 168 (Stirling, 1981) provides suitable versions of these algorithms.

PRECISION

In normal circumstances single precision should be adequate. However, it is possible to convert the routine to double precision by:

(1) Changing *REAL* to *DOUBLE PRECISION* in the declarations;
(2) Changing the real constants in the *DATA* statement to double precision versions;
(3) Changing the right-hand side of the statement function into a suitable double precision function (but note that under standard Fortran 66 there is no equivalent to *FLOAT*, though many compilers offer such routines; alternatively *DBLE(FLOAT(I))* may be used).

RESTRICTIONS

The algorithm assumes that characters, held in 1*H* form, may be assigned from one integer location to another. Attention is also drawn to the final paragraph under the heading *Marks made* earlier, in which the use of the non-Fortran characters '*'*, ';' and ':' is discussed.

The algorithm is not compatible with Fortran 77 because of the different character handling in that language. To facilitate conversion to Fortran 77, all variables used to store characters, and only those variables, are declared explicitly in the *INTEGER* statement.

ACKNOWLEDGEMENT

The author would like to thank the referee for several useful suggestions about the algorithm.

EDITORS' REMARKS

In the original version *IWRITE* was not a parameter of the subroutine, but it has now been included to provide additional flexibility. The corrections noted by Harris (1982) in AS R46 have also been implemented. Attention is drawn to the enhancement of AS R54 due to Dewey (1984) to allow individual points to be identified; this is not incorporated.

The only other changes made are those facilitating the changes to double precision or to Fortran 77 outlined above.

REFERENCES

Dewey, M. E. (1984) Remark AS R54. *Appl. Statist.*, **33**, 370–372.

Harris, R. G. (1982) Remark AS R46. *Appl. Statist.*, **31**, 340.

Sparks, D. N. (1971) Algorithm AS 44. Scatter diagram plotting. *Appl. Statist.*, **20**, 327–331.

Stirling, W. D. (1981) Algorithm AS 168. Scale selection and formatting. *Appl. Statist.*, **30**, 339–344. (See also this book, page 222)

```
      SUBROUTINE SCATPL(A, N, M, ICY, NCY, ICX, IWRITE, NY, NX,
     $  SCALEY, SCALEX, ISTAND, IFAULT)
C
C        ALGORITHM AS 169   APPL. STATIST. (1981) VOL.30, P.345
C
C        PRODUCES A SCATTER PLOT OF ONE VARIABLE AGAINST SEVERAL
C
      DIMENSION ICY(NCY)
      INTEGER IAPOST, IBLANK, ICOLON, ICOMMA, IDASH, IDOT,
     $  IFORM1(19), IFORM2(20), INTCH(11), IOUT(161), ISEMI,
     $  ITWO, MARKCH(5)
      REAL A(N, M), AI, OFFSET, SCALEX(2), SCALEY(2), TEMP,
     $  VALS(20), XMAX, XMIN, XSTEP, XVSTEP, Y, YMAX, YMIN,
     $  YSTEP, YVSTEP, ZERO, HALF, ONEP5, ZFLOAT
C
      DATA ZERO, HALF, ONEP5 /0.0, 0.5, 1.5/
      DATA MAXWID, MAXHT, MAXY /132, 50, 5/
      DATA MPVX, MPVY /10, 5/
      DATA INTCH(1), INTCH(2), INTCH(3), INTCH(4),  INTCH(5),  INTCH(6),
     $     INTCH(7), INTCH(8), INTCH(9), INTCH(10), INTCH(11)
     $  /   1H0,      1H1,      1H2,      1H3,       1H4,
     $      1H5,      1H6,      1H7,      1H8,       1H9,        1H9/
      DATA MARKCH(1), MARKCH(2), MARKCH(3), MARKCH(4), MARKCH(5)
     $  /   1H*,       1H0,       1H+,       1HX,       1H=/
      DATA IBLANK, IDOT, ICOLON, ICOMMA, IAPOST, ISEMI, ITWO, IDASH
     $  /   1H ,  1H.,    1H:,    1H,,    1H',   1H;,  1H2,  1H-/
C
C        FORMATS FOR PRINTING
C
      DATA  IFORM1(1),   IFORM1(2),   IFORM1(3),   IFORM1(4),   IFORM1(5),
     $      IFORM1(6),   IFORM1(7),   IFORM1(8),   IFORM1(9),   IFORM1(10),
     $      IFORM1(11),  IFORM1(12),  IFORM1(13),  IFORM1(14),  IFORM1(15),
     $      IFORM1(16),  IFORM1(17),  IFORM1(18),  IFORM1(19)
     $  /      1H(,         1H1,         1HH,         1H ,         1H,,
     $         1HF,         1H8,         1H.,         1H0,         1H,,
     $         1H1,         1HX,         1H,,         1H1,         1H5,
     $         1H2,         1HA,         1H1,         1H)/
      DATA  IFORM2(1),   IFORM2(2),   IFORM2(3),   IFORM2(4),   IFORM2(5),
     $      IFORM2(6),   IFORM2(7),   IFORM2(8),   IFORM2(9),   IFORM2(10),
     $      IFORM2(11),  IFORM2(12),  IFORM2(13),  IFORM2(14),  IFORM2(15),
     $      IFORM2(16),  IFORM2(17),  IFORM2(18),  IFORM2(19),  IFORM2(20)
     $  /      1H(,         1H1,         1HH,         1H ,         1H,,
     $         1H5,         1HX,         1H,,         1H1,         1H6,
     $         1H(,         1HF,         1H8,         1H.,         1H0,
     $         1H,,         1H2,         1HX,         1H),         1H)/
    1 FORMAT(11X, 1H:, 151A1)
    2 FORMAT(11H TIMES 10**, I3)
    3 FORMAT(7H OFFSET, F10.0)
    4 FORMAT(15X, 10HTIMES 10**, I3)
    5 FORMAT(15X, 6HOFFSET, F10.0)
    6 FORMAT(3X, 16(9X, A1))
```

```
C
      ZFLOAT(I) = FLOAT(I)
C
C         TEST FOR VALID PARAMETERS
C
      IFAULT = 0
      IF (N .LT. 1) IFAULT = IFAULT + 1
      IF (M .LT. 2) IFAULT = IFAULT + 2
      IF (ICX .LT. 1 .OR. ICX .GT. M) IFAULT = IFAULT + 4
      IF (NCY .LE. 0 .OR. NCY .GT. MAXY) IFAULT = IFAULT + 8
      IF (IFAULT .GT. 0) RETURN
      DO 10 I = 1, NCY
      IF (ICY(I) .LT. 1 .OR. ICY(I) .GT. M) GOTO 1016
   10 CONTINUE
C
C         SET PLOT SIZE
C
      NLY = MAXHT - 5
      IF (NLY .GT. NY) NLY = NY
      IF (NLY .LE. MPVY) NLY = MPVY + 1
      NLX = MAXWID - 11
      IF (NLX .GT. NX) NLX = NX
      IF (NLX .LE. MPVX) NLX = MPVX + 1
C
C         CALCULATE HORIZONTAL SCALE
C
      XMIN = SCALEX(1)
      XMAX = SCALEX(2)
      IF (XMAX .GT. XMIN) GOTO 30
      XMIN = A(1, ICX)
      XMAX = XMIN
      IF (N .EQ. 1) GOTO 30
      DO 20 I = 2, N
      AI = A(I, ICX)
      IF (AI .LT. XMIN) XMIN = AI
      IF (AI .GT. XMAX) XMAX = AI
   20 CONTINUE
C
   30 CALL SCALE(XMIN, XMAX, NLX, MPVX, TEMP, XVSTEP,
     $ NXVALS, IRX, IFAIL)
      IF (IFAIL .GT. 0) GOTO 1032
      XMIN = TEMP
      XSTEP = XVSTEP / ZFLOAT(MPVX)
C
C         CALCULATE VERTICAL SCALE
C
      YMIN = SCALEY(1)
      YMAX = SCALEY(2)
      IF (YMAX .GT. YMIN) GOTO 50
      K = ICY(1)
      YMIN = A(1, K)
      YMAX = YMIN
      DO 40 J = 1, NCY
      K = ICY(J)
      DO 40 I = 1, N
      AI = A(I, K)
      IF (AI .LT. YMIN) YMIN = AI
      IF (AI .GT. YMAX) YMAX = AI
   40 CONTINUE
C
   50 CALL SCALE(YMIN, YMAX, NLY, MPVY, TEMP, YVSTEP,
     $ NYVALS, IRY, IFAIL)
      IF (IFAIL .GT. 0) GOTO 1064
      YMIN = TEMP
      YSTEP = YVSTEP / ZFLOAT(MPVY)
C
C         CALCULATE PRINTED VALUES AND SET MARK FOR 2 POINTS
C
      CALL AXIS(YMIN, YVSTEP, NYVALS, 6, IRY, IRPR, OFFSET,
```

```
     $   IFACT, VALS, 20, IFAIL)
         IF (IFAIL .GT. 0) GOTO 1064
         IFORM1(9) = INTCH(IRPR + 1)
         IF (IFACT .NE. 0) WRITE (IWRITE, 2) IFACT
         IF (OFFSET .NE. ZERO) WRITE (IWRITE, 3) OFFSET
         IF (ISTAND .EQ. 0) INTCH(3) = ISEMI
C
C        FOR EACH LINE OF OUTPUT
C
         IPLTED = 0
         DO 140 I = 1, NLY
         IY = NLY - I
         DO 60 IX = 1, NLX
   60 IOUT(IX) = IBLANK
C
C        SCAN DATA FOR POINTS ON CURRENT LINE
C
         DO 120 L = 1, N
         INDX = (A(L, ICX) - XMIN) / XSTEP + ONEP5
         IF (INDX .LT. 1 .OR. INDX .GT. NLX) GOTO 120
         DO 110 J = 1, NCY
         K = ICY(J)
         Y = (A(L, K) - YMIN) / YSTEP
         INDY = Y + HALF
         IF (INDY .NE. IY) GOTO 110
         IPLTED = IPLTED + 1
         IF (IOUT(INDX) .NE. IBLANK) GOTO 80
C
C        SINGLE POINT
C
         IF (ISTAND .EQ. 0) GOTO 70
         IOUT(INDX) = MARKCH(J)
         GOTO 110
   70 IOUT(INDX) = ICOMMA
         IF (INT(Y) .EQ. IY) IOUT(INDX) = IAPOST
         GOTO 110
C
C        MULTIPLE POINTS
C
   80 DO 90 IC = 3, 10
         IF (IOUT(INDX) .EQ. INTCH(IC)) GOTO 100
   90 CONTINUE
         IC = 2
  100 IOUT(INDX) = INTCH(IC + 1)
  110 CONTINUE
  120 CONTINUE
C
C        PRINT LINE
C
         IF (MOD(IY, MPVY) .EQ. 0) GOTO 130
         WRITE (IWRITE, 1) (IOUT(IX), IX = 1, NLX)
         GOTO 140
  130 WRITE (IWRITE, IFORM1) VALS(NYVALS), IDASH, ICOLON,
     $   (IOUT(IX), IX = 1, NLX)
         NYVALS = NYVALS - 1
  140 CONTINUE
C
C        PRINT HORIZONTAL AXIS USING VARIABLE FORMATS
C
         WRITE (IWRITE, 1) (IDOT, I = 1, NLX)
         CALL AXIS(XMIN, XVSTEP, NXVALS, 6, IRX, IRPR, OFFSET,
     $   IFACT, VALS, 20, IFAIL)
         INTCH(3) = ITWO
         IFORM2(15) = INTCH(IRPR + 1)
         IF (IFAIL .GT. 0) GOTO 1032
         WRITE (IWRITE, 6) (ICOLON, I = 1, NXVALS)
         WRITE (IWRITE, IFORM2) (VALS(I), I = 1, NXVALS)
         IF (IFACT .NE. 0) WRITE (IWRITE, 4) IFACT
         IF (OFFSET .NE. 0.0) WRITE (IWRITE, 5) OFFSET
```

```
       IFAULT = IPLTED - N * NCY
       RETURN
C
C         SET ERROR INDICATOR
C
 1064 IFAULT = IFAULT + 32
 1032 IFAULT = IFAULT + 16
 1016 IFAULT = IFAULT + 16
       RETURN
       END
```

Algorithm AS 183

AN EFFICIENT AND PORTABLE PSEUDO-RANDOM NUMBER GENERATOR

By B. A. Wichmann   and   I. D. Hill
National Physical Laboratory,   *Clinical Research Centre,*
Teddington, Middx, TW11 0LW, UK   *Harrow, Middx, HA1 3UJ, UK*

Keywords: Pseudo-random numbers; Efficiency; Portability.

LANGUAGE

Fortran 66 and 77.

DESCRIPTION AND PURPOSE

Schrage (1979) has pointed out the advantages of pseudo-random generators that can be written in a high-level language and produce the same results on any machine. The generator that he presents, however, has the disadvantages: (1) like all simple multiplicative congruential generators, it does not work well at the extremes of the distribution — for any number produced that is less than $5 \cdot 9499 \times 10^{-5}$ the next number will simply be 16,807 times as much, and similarly at the top end; (2) on a 16-bit machine it has to use double precision arithmetic instead of integer arithmetic, which makes it very slow, and also uncertain that rounding errors could not occur.

Our algorithm does not have these difficulties. We claim that it is reasonably short, reasonably fast, machine-independent, easily programmed in any language, and statistically sound. It has a cycle length exceeding $6 \cdot 95 \times 10^{12}$ so that even using 1000 random numbers per second continuously, the sequence would not repeat for over 220 years. Consequently we have tested only small parts of it, consisting of many millions of numbers nevertheless. However, there are theoretical grounds for expecting good results, and the results of the tests we have made have been so satisfactory, that we are prepared to extrapolate our experience and infer that the sequence is satisfactory throughout.

The algorithm produces numbers rectangularly distributed between 0 and 1, excluding the end points but, on some machines, rounding errors might very occasionally produce a value of 0 precisely.

METHOD

Three simple multiplicative congruential generators are used. Each uses a prime number for its modulus and a primitive root for its multiplier, which guarantees a complete cycle. The three results are added, and the fractional part is taken.

It is well known that the sum of n independent rectangular random numbers tends to normality as n increases, but much less well known that the fractional part of such a sum remains rectangular for all values of n. This is most easily seen by considering a simplified case, of two generators that produce numbers of only one digit after the decimal point such as shown in Table 1.

Table 1
Fractional part of $x1 + x2$

Value of $x2$	Value of $x1$									
	0·0	0·1	0·2	0·3	0·4	0·5	0·6	0·7	0·8	0·9
0·0	0·0	0·1	0·2	0·3	0·4	0·5	0·6	0·7	0·8	0·9
0·1	0·1	0·2	0·3	0·4	0·5	0·6	0·7	0·8	0·9	0·0
0·2	0·2	0·3	0·4	0·5	0·6	0·7	0·8	0·9	0·0	0·1
0·3	0·3	0·4	0·5	0·6	0·7	0·8	0·9	0·0	0·1	0·2
0·4	0·4	0·5	0·6	0·7	0·8	0·9	0·0	0·1	0·2	0·3
0·5	0·5	0·6	0·7	0·8	0·9	0·0	0·1	0·2	0·3	0·4
0·6	0·6	0·7	0·8	0·9	0·0	0·1	0·2	0·3	0·4	0·5
0·7	0·7	0·8	0·9	0·0	0·1	0·2	0·3	0·4	0·5	0·6
0·8	0·8	0·9	0·0	0·1	0·2	0·3	0·4	0·5	0·6	0·7
0·9	0·9	0·0	0·1	0·2	0·3	0·4	0·5	0·6	0·7	0·8

It is immediately clear from this table that if $x2$ is equally likely to take any of its ten possible values, then so is the fractional part of $x1 + x2$, provided that $x1$ and $x2$ are independent, whatever the value of $x1$ may be. It therefore remains so, whatever the distribution of $x1$. It follows that where we have n such variables, the joint distribution of $n - 1$ of them is immaterial provided that the other one is random rectangular and independent.

In practice, however, no pseudo-random generator is perfect, and adding several in this way leads to the imperfections of each being 'ironed out' by the others.

Each generator has a period one unit less than the value of its prime modulus. These periods are therefore even numbers and necessarily have 2 as a common factor, but the values used have no other common factor, thus producing a combined cycle-length equal to one quarter of the product of the individual cycle-lengths. Because of this common factor, the three generators are not quite independent of each other, but the tests made indicate that the results are excellent in spite of this.

STRUCTURE

REAL FUNCTION RANDOM (L)

Formal parameter

L	Integer	input:	a dummy argument. Any value will do.

Common
COMMON /RAND/ IX, IY, IZ

These three integers must be set before the first entry, to values within the range 1 − 30,000, preferably chosen at random. Where the machine has facilities for accessing the date and the time, suitable values may be derived from these. If the values are printed before the first use of the algorithm, they are available for a repeat run if required. If they are printed after the last use, they give suitable values for entry on another occasion with the certainty of a different sequence.

Once set, these numbers should not be changed, other than by calls of the algorithm.

These values could be passed as arguments instead of through *COMMON*, but on many machines using *COMMON* will be faster.

RESTRICTIONS

Integer arithmetic up to 30,323 is required.

It is assumed, without testing, that the three integer values are within the correct range. It does not seem worth testing at every entry, because if the values are suitable at the first entry they necessarily remain so at subsequent entries.

MODIFICATION

A modified, and simpler, version is mentioned in a comment. This requires integer arithmetic up to 5,212,632 and may be slightly faster on some machines; however, it is much slower on other machines, and should not be used without testing for speed.

TIME

On a PDP-11 computer each call of the algorithm takes 0·85 ms. This compares with 4·4 ms. for Schrage's algorithm (in its double-precision version that is necessary for a 16-bit machine), and with 1·3 ms. for the generator included in the *GLIM* program.

TESTS

Results of the serial test, poker test, coupon collector's test and runs-up-and-down test performed on this generator are given in Wichmann and Hill (1982). This reference also contains further discussion. We are grateful to the referee for additional, very thorough, testing.

PRECISION

The initial line may be changed from *REAL FUNCTION* to *DOUBLE PRE-CISION FUNCTION* if a result in double precision form is desired. There would be no point, however, in changing the internal working of this algorithm, so the usual devices to make such a change easy are not incorporated.

ASSOCIATED ALGORITHM

To produce pseudo-random normal deviates, the results of this algorithm may be used as input to AS 111 (Beasley and Springer, 1977), but precautions are then necessary against an exactly zero result.

ACKNOWLEDGEMENT

Thanks are due to A. I. McLeod for demonstrating that, on some machines, an exact zero can very occasionally occur.

EDITORS' REMARKS

Apart from the addition of the word *REAL* at the start, the algorithm remains as first published. Some necessary corrections have been made to the introductory text.

REFERENCES

Beasley, J. D. and Springer, S. G. (1977) Algorithm AS 111. The percentage points of the normal distribution. *Appl. Statist.*, **26**, 118–121. (See also this book, page 188)

Schrage, L. (1979) A more portable Fortran random number generator. *ACM Trans. Math. Softw.*, **5**, 132–138.

Wichmann, B. A. and Hill, I. D. (1982) *A pseudo-random number generator.* NPL Report DITC 6/82.

```
      REAL FUNCTION RANDOM(L)
C
C         ALGORITHM AS 183   APPL. STATIST. (1982) VOL.31, P.188
C
C         RETURNS A PSEUDO-RANDOM NUMBER RECTANGULARLY DISTRIBUTED
C         BETWEEN 0 AND 1.
C
C         IX, IY AND IZ SHOULD BE SET TO INTEGER VALUES BETWEEN
C         1 AND 30000 BEFORE FIRST ENTRY.
C
C         INTEGER ARITHMETIC UP TO 30323 IS REQUIRED.
C
      COMMON /RAND/ IX, IY, IZ
      IX = 171 * MOD(IX, 177) - 2  * (IX / 177)
      IY = 172 * MOD(IY, 176) - 35 * (IY / 176)
      IZ = 170 * MOD(IZ, 178) - 63 * (IZ / 178)
C
      IF (IX .LT. 0) IX = IX + 30269
      IF (IY .LT. 0) IY = IY + 30307
      IF (IZ .LT. 0) IZ = IZ + 30323
```

```
C
C              IF INTEGER ARITHMETIC UP TO 5212632 IS AVAILABLE,
C              THE PRECEDING 6 STATEMENTS MAY BE REPLACED BY
C
C              IX = MOD(171 * IX, 30269)
C              IY = MOD(172 * IY, 30307)
C              IZ = MOD(170 * IZ, 30323)
C
C              ON SOME MACHINES, THIS MAY SLIGHTLY INCREASE
C              THE SPEED. THE RESULTS WILL BE IDENTICAL.
C
       RANDOM = AMOD(FLOAT(IX) / 30269.0 + FLOAT(IY) / 30307.0 +
      $              FLOAT(IZ) / 30323.0, 1.0)
       RETURN
       END
```

Algorithm ACM 291

LOGARITHM OF GAMMA FUNCTION
By M. C. Pike[†] and I. D. Hill[‡]
MRC Statistical Research Unit

Keywords: Gamma function; Stirling's formula.

LANGUAGES
Algol 60, and Fortran 66 and 77.

DESCRIPTION AND PURPOSE
The algorithm evaluates the natural logarithm of $\Gamma(x)$ for all $x > 0$.

NUMERICAL METHOD
If x is at least $7 \cdot 0$ Stirling's formula is used. Otherwise Stirling's formula is applied to $7 \cdot 0$ plus the fractional part of x. The result is then determined from the recurrence relation

$$\Gamma(n + 1) = n \, \Gamma(n)$$

STRUCTURE
real procedure *loggamma (x, ifault)*

Formal parameters

x	Real	value:	value of x for which $\ln \Gamma(x)$ is required. Must exceed zero.
ifault	Integer	output:	1 if $x \leqslant 0$; 0 otherwise.

ACCURACY
The result is accurate to at least 8 significant digits (decimal), except where the result is less than $0 \cdot 1$ when the accuracy is to at least 8 digits past the decimal point, provided that the computer can give such accuracy.

[†]*Present address:* Cancer Epidemiology and Clinical Trials Unit, Radcliffe Infirmary, Oxford, UK.

[‡]*Present address:* Clinical Research Centre, Harrow, Middx. HA1 3UJ, UK.

This algorithm is taken, with permission, from the ACM series because a number of *Applied Statistics* algorithms use it. The original version has been modified to incorporate Miss Hoare's (1968) remark, and further improved to take advantage of the latest definition of Algol 60. A Fortran translation is given for compatibility with other algorithms in this book. A fault parameter has been added to ensure that only positive values of x are handled.

REFERENCES

Hoare, M. R. (1968) Remark on Algorithm 291. *Comm. ACM.*, **11**, 14.
Pike, M. C. and Hill, I. D. (1966) Algorithm 291. *Comm. ACM.*, **9**, 684.

```
real procedure loggamma(x, ifault);

comment Algorithm ACM 291 Comm. ACM. (1966) Vol.9, p.684;

value x; real x; integer ifault;

comment evaluates natural logarithm of gamma(x), for x > 0;

if x ≤ 0.0 then
    begin
    loggamma := 0.0; ifault := 1
    end
else
    begin real f, z;
    own real a1, a2, a3, a4, a5;
    if a1 < 0.5 then
        begin

        comment at first entry, set constants.
        a1 is half the natural logarithm of 2pi;

        a1 := ln(8.0 × arctan(1.0)) / 2.0;
        a2 := 1.0 / 1680.0; a3 := 1.0 / 1260.0;
        a4 := 1.0 / 360.0; a5 := 1.0 / 12.0
        end;
    ifault := 0;
    if x < 7.0 then
        begin
        f := x;
        for x := x + 1.0 while x < 7.0 do f := f × x;
        f := − ln(f)
        end
```

```
          else f := 0.0;
          z := 1.0 / x ↑ 2;
          loggamma := f + (x − 0.5) × ln(x) − x + a1 +
              (((−a2 × z + a3) × z − a4) × z + a5) / x
          end loggamma
```

Copyright 1966, Association for Computing Machinery, Inc., reprinted by permission.

Fortran version of Algorithm ACM 291

STRUCTURE

REAL FUNCTION ALOGAM (X, IFAULT)

Formal parameters
X and *IFAULT* serve identical purposes to *x* and *ifault* in the Algol version.

PRECISION

For a double precision version, change *REAL FUNCTION* to *DOUBLE PRE-CISION FUNCTION*, *REAL* to *DOUBLE PRECISION*, *ALOG* to *DLOG* and give double precision versions of the constants in the *DATA* statements.

```
          REAL FUNCTION ALOGAM(X, IFAULT)
C
C             ALGORITHM ACM 291, COMM. ACM. (1966) VOL.9, P.684
C
C             EVALUATES NATURAL LOGARITHM OF GAMMA(X)
C             FOR X GREATER THAN ZERO
C
          REAL A1, A2, A3, A4, A5, F, X, Y, Z, ZLOG,
     $    HALF, ZERO, ONE, SEVEN
C
C             THE FOLLOWING CONSTANTS ARE ALOG(2PI)/2,
C             1/1680, 1/1260, 1/360 AND 1/12
C
          DATA A1, A2, A3, A4, A5
     $    /0.91893 85332 04673, 0.00059 52380 95238,
     $    0.00079 36507 93651, 0.00277 77777 77778,
     $    0.08333 33333 33333/
          DATA HALF, ZERO, ONE, SEVEN /0.5, 0.0, 1.0, 7.0/
C
          ZLOG(F) = ALOG(F)
C
          ALOGAM = ZERO
          IFAULT = 1
          IF (X .LE. ZERO) RETURN
          IFAULT = 0
          Y = X
          F = ZERO
          IF (Y .GE. SEVEN) GOTO 30
          F = Y
```

```
10 Y = Y + ONE
   IF (Y .GE. SEVEN) GOTO 20
   F = F * Y
   GOTO 10
20 F = -ZLOG(F)
30 Z = ONE / (Y * Y)
   ALOGAM = F + (Y - HALF) * ZLOG(Y) - Y + A1
 $   + (((-A2 * Z + A3) * Z - A4) * Z + A5) / Y
   RETURN
   END
```

Index of statistical algorithms

The following is an index to all the algorithms published in the *Applied Statistics* series up to and including the final issue of 1984. The algorithm reference number and the title of the algorithm are tabulated together with the Journal references in the form 'Volume, Page' both for the original publication and any Remarks or Corrections which have appeared subsequently. Also listed are those other algorithms which are required or for which that algorithm is needed, as well as any other related algorithms in the series. For those algorithms included in this book, the reference given in the column headed 'Vol, Page' is to the page in this book and not to the original publication in *Applied Statistics*. Thus, within this list, the algorithms contained in this book can readily be identified as their entries in the 'Vol, Page' column lack the volume number and the comma. For these algorithms no entries are provided in the column headed 'Remarks etc.', since all Remarks and Corrections published up to the end of 1984 have been taken into account in the preparation of the versions included in this book; references to such Remarks can be found in the appropriate description as presented earlier in the book.

Indexes of this type have appeared on occasions in *Applied Statistics*, the latest being in 1981. Those indexes have always ordered the algorithms by subject area according to the *ISI* classification scheme which was published in *Applied Statistics*, Vol. 25, 1976, pp. 100–101. Such classification is difficult because many of the algorithms could equally well be classified under several different headings. In view of this and the provision of a Keyword Index, the index below is presented in numerical order.

ALG. NO.	VOL, PAGE	TITLE	CALLS ALGS.	CALLED BY ALGS.	RELATED ALGS.	REMARKS ETC.
AS 1	17,180	Simulating multidimensional arrays in one dimension		AS 18 AS 19 AS 23	AS 28 AS 39	18,116
AS 2	17,186	The Normal integral		AS 4 AS 5 AS 50	AS 17 AS 24 AS 66 AS 195	18,299
AS 3	38	The integral of Student's t-distribution			AS 5 AS 27 AS 184	
AS 4	17,190	An auxiliary function for distribution integrals	AS 2	AS 5	AS 76	18,118 19,204 22,428
AS 5	40	The integral of the non-central t-distribution	AS 66 AS 76 ACM291		AS 3 AS 76 AS 184	
AS 6	43	Triangular decomposition of a triangular matrix		AS 7 AS 53 AS 84 AS 122 AS 125 AS 139 AS 162 AS 194 AS 196	AS 46 AS 75 AS 163 AS 191	
AS 7	46	Inversion of a positive semi-definite symmetric matrix	AS 6	AS 84 AS 122 AS 125 AS 139 AS 162 AS 174 AS 194 AS 196	AS 37 AS 195	
AS 8	17,277	Main effects from a multi-way table			AS 9	
AS 9	17,279	Construction of additive table			AS 8	
AS 10	17,283	The use of orthogonal polynomials		AS 42 AS 166		20,117 20,216
AS 11	17,287	Normalizing a symmetric matrix			AS 12	
AS 12	17,289	Sums of squares and products matrix			AS 11 AS 41 AS 50 AS 52 AS 72 AS 95 AS 110	
AS 13	49	Minimum spanning tree			AS 14 AS 15 AS 40 AS 102	

ALG. NO.	VOL, PAGE	TITLE	CALLS ALGS.	CALLED BY ALGS.	RELATED ALGS.	REMARKS ETC.
AS 14	53	Printing the minimum spanning tree			AS 13 AS 15	
AS 15	56	Single linkage cluster analysis			AS 13 AS 14	
AS 16	18,110	Maximum likelihood estimation from grouped and censored normal data	AS 17		AS 95 AS 138 AS 139	26,122
AS 17	18,115	The reciprocal of Mills's ratio		AS 16	AS 2 AS 66 AS 138 AS 139	
AS 18	18,197	Evaluation of marginal means	AS 1		AS 19	
AS 19	18,199	Analysis of variance for a factorial table	AS 1		AS 18	
AS 20	18,203	The efficient formation of a triangular array with restricted storage for data				
AS 21	18,206	Scale selection for computer plots			AS 30 AS 44 AS 96	20,118 23,248
AS 22	18,283	The interaction algorithm			AS 23	
AS 23	18,287	Calculation of effects	AS 1		AS 22	
AS 24	18,290	From Normal integral to deviate			AS 2 AS 70	
AS 25	18,294	Classification of means from analysis of variance				
AS 26	19,111	Ranking an array of numbers	CJ 26			22,133
AS 27	19,113	The integral of Student's t-distribution			AS 3	
AS 28	19,115	Transposing multiway structures			AS 1	
AS 29	19,190	The runs up and down test			AS 53 AS 98 AS 157	25,193
AS 30	65	Half Normal plotting			AS 21	
AS 31	19,197	Operating characteristic and average sample size for binomial sequential sampling			AS 67	
AS 32	19,285	The incomplete gamma integral	ACM291	AS 91 AS 184	AS 147 AS 187	
AS 33	19,287	Calculation of hyper-geometric sample sizes			AS 59	

ALG. NO.	VOL, PAGE	TITLE	CALLS ALGS.	CALLED BY ALGS.	RELATED ALGS.	REMARKS ETC.
AS 34	19,290	Sequential inversion of band matrices				
AS 35	20, 99	Probabilities derived from finite populations	ACM347		AS 55 AS 56 AS 62 AS 93 AS 142	20,346 21,352 26,221
AS 36	20,105	Exact confidence limits for the odds ratio in a 2 x 2 table			AS 115	30, 73
AS 37	20,111	Inversion of a symmetric matrix			AS 7 AS 38 AS 141 AS 167 AS 178	23,100
AS 38	20,112	Best subset search			AS 37	
AS 39	20,115	Arrays with a variable number of dimensions			AS 1	
AS 40	20,204	Updating a minimum spanning tree			AS 13	
AS 41	70	Updating the sample mean and dispersion matrix			AS 12 AS 52 AS 72	
AS 42	20,209	The use of orthogonal polynomials with equal x-values	AS 10			
AS 43	20,213	Variable format in Fortran				20,346
AS 44	20,327	Scatter diagram plotting			AS 21 AS 73 AS 96 AS 169	23,248
AS 45	74	Histogram plotting				
AS 46	20,335	Gram-Schmidt orthogonalization			AS 6 AS 75 AS 79	
AS 47	79	Function minimization using a simplex procedure				
AS 48	21, 97	Uncertainty function for a binary sequence			AS 49	
AS 49	21,100	Autocorrelation function for a binary sequence			AS 48	
AS 50	21,103	Tests of fit for a one-hit vs two-hit curve	AS 2		AS 12	
AS 51	88	Log-linear fit for contingency tables		AS 185	AS 160 AS 207	

ALG. NO.	VOL, PAGE	TITLE	CALLS ALGS.	CALLED BY ALGS.	RELATED ALGS.	REMARKS ETC.
AS 68	23, 87	A program for estimating the parameters of the truncated negative binomial distribution				
AS 69	23, 92	Knox test for space-time clustering in epidemiology				
AS 70	23, 96	The percentage points of the Normal distribution		AS 87 AS 91 AS 190 AS 195	AS 24 AS 100 AS 111	
AS 71	23, 98	The upper tail probabilities of Kendall's tau	AS 66		AS 54	
AS 72	23,234	Computing mean vectors and dispersion matrices in multivariate analysis of variance			AS 12 AS 41 AS 52	
AS 73	23,238	Cross-spectrum smoothing via the finite Fourier transform			AS 44	30,354
AS 74	23,244	L1-norm of a straight line			AS 108 AS 110 AS 132	25, 96
AS 75	130	Basic procedures for large, sparse or weighted linear least squares problems			AS 6 AS 46 AS 79 AS 154 AS 163 AS 164	
AS 76	145	An integral useful in calculating non-central t and bivariate Normal probabilities			AS 4 AS 5	
AS 77	23,458	Null distribution of the largest root statistic				
AS 78	23,466	The mediancentre			AS 143	24,390
AS 79	23,470	Gram-Schmidt regression			AS 46 AS 75	
AS 80	24,144	Spherical statistics	BIT 3,205			
AS 81	24,147	Circular statistics	BIT 3,205		AS 86	
AS 82	24,150	The determinant of an orthogonal matrix				
AS 83	149	Complex discrete fast Fourier transform		AS 97 AS 117		
AS 84	24,262	Measures of multivariate skewness and kurtosis	AS 6 AS 7			

ALG. NO.	VOL, PAGE	TITLE	CALLS ALGS.	CALLED BY ALGS.	RELATED ALGS.	REMARKS ETC.
AS 154	206	An algorithm for exact maximum likelihood estimation of autoregressive-moving average models by means of Kalman filtering		AS 182	AS 75 AS 191 AS 197	
AS 155	29,323	The distribution of a linear combination of chi-squared random variables			AS 106 AS 204	33,366
AS 156	29,334	Combining two component designs to form a row-and-column design				
AS 157	30, 81	The runs-up and runs-down tests			AS 29 AS 98	
AS 158	30, 85	Calculation of the probabilities {p(l,k)} for the simply ordered alternative		AS 198	AS 90 AS 122 AS 123	
AS 159	30, 91	An efficient method of generating random R*C tables with given row and column totals			AS 144 AS 205	
AS 160	30, 97	Partial and marginal association in multidimensional contingency tables			AS 51	
AS 161	30,182	Critical regions of an unconditional non-randomized test of homogeneity in 2*2 contingency tables				
AS 162	30,190	Multivariate conditional logistic analysis of stratum-matched case-control studies	AS 6 AS 7		AS 196	33,123
AS 163	30,198	A Givens algorithm for moving from one linear model to another without going back to the data			AS 6 AS 75 AS 164	
AS 164	30,204	Least squares subject to linear constraints			AS 75 AS 163	30,357
AS 165	30,313	An algorithm to construct a discriminant function in Fortran for categorical data				
AS 166	30,325	Generation of polynomial contrasts for incomplete factorial designs with quantitative levels	AS 10		AS 167	

List of routine names

This is an alphabetical list of the names of all externally called routines and **COMMON** blocks which have appeared in the algorithms published in the *Applied Statistics* series up to and including the final issue of 1984. This list is provided primarily for authors of software who might wish to ensure that they are choosing names for their routines which will not clash with any already in the *Applied Statistics* series, since it is prudent to avoid the duplication of routine names in a software library. Where an algorithm consists of more than one routine, the various routines are differentiated from one another, both in the Journal and in this list, by the notation AS n.1, AS n.2, etc. Routine names and references followed by a '∗' indicate that the routine named must be supplied by the user in accordance with the specification given in the published Description; those followed by a '+' are the names of Fortran versions of Algol 60 routines written specially for this book. Entry names which are enclosed within solidi refer to **COMMON** block names.

Routine Name	Algorithm Reference	Routine Name	Algorithm Reference
ACOS *	AS 90.1*	checkset	AS 1.1
ADDEFF	AS 8	CHFOR	AS 117.2
address	AS 1.3	CHI *	AS 50.6*
ADDTAB	AS 9	CHI	AS 170.2
ADJUST	AS 51.2	CHISQL	AS 170.1
AJV	AS 100.1	CHISQN	AS 170
ALGFAC	AS 196.2	CHOL	AS 6
ALIAS	AS 164.4	CHREV	AS 117.3
ALLNR	AS 88	CHRFT	AS 117.1
ALLOC	AS 140.2	CHRP	AS 117.5
ALNFAC	AS 177.2	CHYPER	AS 152
ALNORM	AS 66	CIM	AS 25
ALOGAM *	AS 147.1*	circstat	AS 81
AMALGM	AS 149	CLUSTR	AS 58
ANALYZ +	AS 75.5+	COLLAP	AS 51.1
analyze	AS 75.5	COMB	AS 173.3
AOV	AS 19	COMBIN	AS 161.1
APRANK	AS 118	COMBO	AS 160.2
arccos *	AS 80.2*	COMBRC	AS 156
arccos *	AS 81.2*	COMICS	AS 201
ASCEND *	AS 101.2*	CONDJ	AS 146.1
ASIN *	AS 158.6*	CONF	AS 160.1
ASRCH	AS 185	CONFND +	AS 75.2+
AXIS	AS 168.1	confound	AS 75.2
/B/	AS 159	CONINT	AS 101
BAB	AS 199	/CONS/	AS 128
BANINV	AS 34	CONVRT	AS 57.1
BASE2	AS 146.2	CORREC	AS 177.4
BAUER	AS 175.1	count	AS 35.1
BBL	AS 189.3	COV	AS 128.3
BBME	AS 189.1	COVMAT	AS 128
BBML	AS 189	COVUP	AS 72
BETAIN	AS 63	CPYCON	AS 185.5
/BETCOM/	AS 134	CPYLR	AS 185.6
BET1GT	AS 134.1	CRSWOP *	AS 113.4*
BINAUTOC	AS 49	CRTRAN *	AS 113.3*
BININFO	AS 48	CSAD	AS 73
BISORT	AS 65.1	CURVE	AS 95
BIVCNT	AS 151	CVSUR1	AS 125
BIVNOR *	AS 195.2*	CVSUR2	AS 125.1
/BKHIT/	AS 50	DATUM *	AS 12.1*
/BK1/	AS 36	DEGREF	AS 185.3
/BLKT/	AS 50	DEN	AS 137.1
BLUS	AS 104	DENEST	AS 176
BMIX	AS 123	DER	AS 128.1
BREJ	AS 194	DESEND *	AS 101.3*
BSUB	AS 164.1	DESMAT	AS 173
Burman	AS 31.1	DETQ	AS 82
BWORTH	AS 42	DEVIAT *	AS 95.3*
centnorm *	AS 204.1*	DIFF	AS 101.1
CHASE	AS 198.2	DIGAMA	AS 103

Routine Name	Algorithm Reference	Routine Name	Algorithm Reference
DIGAMI	AS 187	GIVENC	AS 164
DINT	AS 193.3	GLLM	AS 207
DIST	AS 206.2	GMANWIT	AS 55
divide *	AS 104.1*	GMSMOD	AS 46
DNINT	AS 193.2	GRADSOL	AS R52
DOWN	AS 94.1	GSCALE	AS 93
DSSP	AS 41	GSREG	AS 79
DTARMA	AS 191.1	GSWEEP	AS 178
effects	AS 23.1	GTRANS	AS 163
eigenvectors *	AS 104.2*	G05AAF *	AS 159.1*
EM	AS 138	HERMIT	AS 87.1
ENTNGL	AS 166	HISTGM	AS 45
ENUM	AS 205	HIT	AS 50
EODSEQ	AS 188	HNPLOT	AS 30
EODSRT	AS 188.1	HOWARD	AS 196.1
EPOWTH	AS 129	HYP *	AS 80.1*
EVAL	AS 160.3	HYP *	AS 81.1*
EVAL *	AS 205.1*	HYPER	AS 59
EVDER	AS 202.2	HYPERG	AS 33
EXACT	AS 142	ICLAS	AS 165.3
EXTR	AS 133	ICNT	AS 65.3
FAC	AS 146.3	ICOUNT	AS R22
FACSYM	AS 175	IMPLY	AS 93.4
FACT	AS 158.5	INCLUD +	AS 75.1+
FASTF	AS 83.1	include	AS 75.1
FASTG	AS 83.2	include2	AS R17
FFT	AS 186.1	INCLU2	AS 154.3
FISHUP	AS 129.4	INDLOW	AS 130.3
FIT	AS 167	INIT	AS 140.1
FLIKAM	AS 197	INIT	AS 177.1
FLOG	AS 33.1	INITL	AS 95.1
FNE	AS 134.3	INOUT	AS 115.1
FOLD *	AS 194.2*	INTER	AS 173.2
FORKAL	AS 182	INVERT *	AS 195.3*
FORRT	AS 97.1	INVEST	AS 105
FRQADD	AS 93.3	IOR	AS 65.2
FUN	AS 36.2	IPOWR	AS 165.1
FUNC *	AS 95.2*	ITERA	AS 115.2
FUNCT	AS 4	JFTABT	AS 119
F1	AS 158.2	JKKN	AS 148
F2	AS 158.4	JNSN	AS 99
GAMAIN	AS 32	JOB *	AS 88.1*
GAML *	AS 90.2*	JOB *	AS 179.1*
GAMMDS	AS 147	JOINP	AS 146
GAUINV	AS 70	KALFOR	AS 154.2
GDER	AS 189.4	KARMA	AS 154.1
get	AS 56.1	KKEY	AS 165
get logrank	AS 56	KMNS	AS 136
GETONE	AS 142.1	KNOX	AS 69
GETVAL *	AS 119.1*	KTHNN	AS 202.1
GETVAL *	AS 131.1*	LFNORM	AS 135

Routine Name	Algorithm Reference	Routine Name	Algorithm Reference
LIKE	AS 160.5	PEARSN	AS 192
LIMITS	AS 36	Pearson12	AS 56.2
LINCOM	AS 175.3	PERMUT	AS 179
LOCALM *	AS 133.2*	PHI	AS 195.1
LOGCCH	AS 196	pivot	AS 37
LOGCCS	AS 162	PIVOT	AS 167.1
LOGFIT	AS 160.6	PLOT6	AS 61
LOGLIN	AS 51	POLLY	AS 94
LONESL	AS 74	POLRT *	AS 87.2*
LOWEFF	AS 185.1	POLY	AS 181.2
LPEST	AS 110	POLYCR	AS 87
LRCHSQ	AS 185.2	POWTH	AS 129.2
LRVT	AS 60.2	PPCHI2	AS 91
LS2WAY	AS 120	PPND	AS 111
MANEFF	AS 173.1	PRHO	AS 89
matmap	AS 28.2	Primtree	AS 13
MAX	AS 50.2	PRM	AS 129.1
Maxlicens	AS 16	PRNCST	AS 5
MCHOL	AS 191.2	PROB	AS 90
means	AS 18	PROB	AS 184
MEDIAN	AS 78	PROBA	AS 115.3
MEDIAN	AS 143	PROBS	AS 158
mintreeprint	AS 14	PROBS	AS 198
mixture	AS 203	PROBST	AS 3
MLE	AS 194.1	PRTAUS	AS 71
MODLIK	AS 202	PRTREE +	AS 13+
MOM	AS 99.3	PRTRNG	AS 190
MOMNTS	AS 52	PR1	AS 158.1
MSAE	AS 108	PR2	AS 158.3
MSTAT	AS 130	QDIST	AS 106
MSTUPD	AS 40	qf	AS 155
MTP +	AS 14+	QTRAN	AS 136.2
MULMAX	AS 145	QTRNG	AS 190.1
MULNOR	AS 195	QTRNG0	AS 190.2
MWP	AS 50.3	R	AS 39.2
NELMIN	AS 47	RAND	AS 53.1
nextstep	AS 38	/RAND/	AS 183
NONPAR	AS 174	RANDOM *	AS 53.2*
NORMAL	AS 2	RANDOM	AS 183
NORMAL *	AS 127.1*	RANF *	AS 134.4*
nptss	AS 124	RANF *	AS 137.2*
NSCOR1	AS 177	rank	AS 26
NSCOR2	AS 177.3	RBINOM	AS 67
OPTRA	AS 136.1	RBO	AS 186
ORTHON	AS 10	RCONT	AS 144
OUTFRG *	AS 188.2*	RCONT2	AS 159
OUTPUT	AS 50.4	realcount	AS 35.2
PAN	AS 153	RECURS	AS 175.2
PANDQ	AS 114	REGRES	AS 139
PAPRX	AS 198.1	REGRES	AS 154.4
PAV	AS 206.1	regress	AS 75.4

Routine Name	Algorithm Reference	Routine Name	Algorithm Reference
REGRSS +	AS 75.4+	SOL	AS 36.1
REJECT	AS 129.3	SOLVE	AS 125.2
RESET	AS 160.4	SORTR *	AS 130.4*
RESIDU	AS 175.4	SPEC	AS 98
REVERS	AS 172.1	SPECT	AS 193
REVRT	AS 97.2	sphstat	AS 80
RMills	AS 17	SPLIT	AS 151.1
RMILLS	AS 138.1	SQUANT	AS 130.2
RNGPI	AS 126	SRCH	AS 161.4
RNORM *	AS 50.5*	SSCOMP	AS 164.2
RNORTM	AS 127	SSDCMP +	AS 75.3+
ROOT	AS 92	SS decomp	AS 75.3
ROY TST	AS 77	SSPCOR	AS 11
RUBEN	AS 204	SSQPRO	AS 12
runsud	AS 29	STARMA	AS 154
RWNORM	AS 128.4	START1	AS 93.1
SARMAS	AS 191	START2	AS 93.2
SBFIT	AS 99.2	STUDENT	AS 27
SCALE	AS 21	SUBCHL	AS R44
SCALE	AS 96	SUBGRP	AS 25.1
SCALE	AS 168	SUBINV	AS R44.1
scan	AS 1.4	SUFIT	AS 99.1
SCATPL	AS 44	SUMSQ	AS 200
SCATPL	AS 169	SWOP	AS 113.2
SCORE	AS R22.2	SYMINV	AS 7
SCOUNT	AS 150	T	AS 39.1
SCRAG *	AS 97.3*	T	AS 50.1
SCRAM	AS 83.3	TABLE	AS 161
SCRAM	AS 83.4	tabmap	AS 28.3
SCREEN	AS 160	TABPDF	AS 112
SDSDOT *	AS 175.5*	TABWRT	AS 57
SEQUAL	AS 107	TBYTCI	AS 115
sequential	AS 31.2	TDIAG	AS 60.1
SET	AS 189.2	/TEMPRY/	AS 159
setup	AS 1.2	TETRA	AS 116
SETWT	AS 117.4	TFN	AS 76
SFINT	AS 65	THTHT *	AS 148.1*
SFREQ	AS 54	TMAX	AS 161.3
signtest	AS 85	TOPT	AS 134.2
SIMDO	AS 172	TPSHV	AS 171
SIMLP	AS 132	TRANS	AS 98.1
SIMPAT	AS 137	transpose	AS 28.1
SIMPLY	AS R22.1	TRIGAM	AS 121
singlelinkage	AS 15	TRIMSD	AS 180
SINV *	AS 120.1*	TRISET	AS 20
SINV	AS 141	TRNGBN	AS 68
SKEWKT	AS 84	TRNSFR	AS 113.1
SLINK +	AS 15+	TRUNC	AS 165.2
SMAVG	AS 130.1	TWACF	AS 197.1
SMOOTH	AS 206	TWIDL	AS 162.1
SNV	AS 100.2	UDIST	AS 62

Routine Name	Algorithm Reference	Routine Name	Algorithm Reference
UDRUNS	AS 157	WEIGHT	AS 122
UFTABH	AS 131	WEXT	AS 181
ULTRAM	AS 102	WSHRT	AS 53
UMPUAL	AS 161.2	XFROMP	AS 24
UNPACK	AS 139.1	XINBTA	AS 64
UPDATE	AS 135.1	XINBTA	AS 109
UPDATE	AS 163.1	XLIKE	AS 140.3
UVNO *	AS 144.1*	XNOR *	AS 122.1*
VAR	AS 128.2	XSTAT	AS 112.1
VAR	AS 164.3	XTAIL	AS 185.4
VFOR	AS 43	YATES	AS 22
vonmises	AS 86	Yates	AS 23.2
VPROD	AS 193.1	YL	AS 133.1
WCOEF	AS 181.1	ZERO *	AS 194.3*

Keyword Index

Starting with **AS** 53, a small number of keywords have been provided by the authors of each algorithm as it appeared in *Applied Statistics*. Those algorithms appearing in this book which did not originally have keywords have had appropriate ones supplied in collaboration with the authors. To provide a comprehensive index, keywords have also been assigned to all other algorithms which did not originally have them, but not in collaboration with the authors.

In this index, each word in a 'keyword phrase' is indexed. A somewhat unusual layout has been used so that the word being indexed appears in the centre of the page, alphabetically and always beginning with a capital letter, and is flanked by the remainder of the phrase. Thus, the keyword 'tail area' appears under both 'Area' and 'Tail', as 'tail Area' and 'Tail area' respectively.

The algorithm number is given on the extreme left of the page, and its reference on the extreme right. References are to the *Applied Statistics* journal in the form 'volume, page', except that, for algorithms included in this book, no volume number is given, and the reference is to the page number in this book.

It should be noted that no attempt has been made to rationalize the keywords, and they appear exactly as in the Journal, so that similar topics may not all be grouped together. For example 'Estimate', 'Estimation' and 'Estimator' will merit separate entries in the index, but could well relate to similar material.

Index to Chapters 1 and 2